D0367983

Dividing the Waters: Food Security, Ecosystem Health, and the New Politics of Scarcity

SANDRA POSTEL

Jane A. Peterson, *Editor*

WORLDWATCH PAPER 132
September 1996

THE WORLDWATCH INSTITUTE is an independent, nonprofit environmental research organization in Washington, D.C. Its mission is to foster a sustainable society in which human needs are met in ways that do not threaten the health of the natural environment or future generations. To this end, the Institute conducts interdisciplinary research on emerging global issues, the results of which are published and disseminated to decisionmakers and the media.

FINANCIAL SUPPORT is provided by Carolyn Foundation, the Nathan Cummings Foundation, the Geraldine R. Dodge Foundation, The Ford Foundation, the George Gund Foundation, The William and Flora Hewlett Foundation, W. Alton Jones Foundation, John D. and Catherine T. MacArthur Foundation, Andrew W. Mellon Foundation, The Curtis and Edith Munson Foundation, The Pew Charitable Trusts, Lynn R. and Karl E. Prickett Fund, Rasmussen Foundation, Rockefeller Brothers Fund, Rockefeller Financial Services, Surdna Foundation, Turner Foundation, U.N. Population Fund, Wallace Genetic Foundation, Wallace Global Fund, Weeden Foundation, and the Winslow Foundation.

PUBLICATIONS of the Institute include the annual *State of the World*, which is now published in 27 languages; *Vital Signs*, an annual compendium of global trends that are shaping our future; the *Environmental Alert* book series; *World Watch* magazine; and the Worldwatch Papers. For more information on Worldwatch publications, write: Worldwatch Institute, 1776 Massachusetts Ave., NW, Washington, DC 20036; or fax 202-296-7365; or see back pages.

THE WORLDWATCH PAPERS provide in-depth, quantitative and qualitative analysis of the major issues affecting prospects for a sustainable society. The Papers are written by members of the Worldwatch Institute research staff and reviewed by experts in the field. Published in five languages, they have been used as concise and authoritative references by governments, nongovernmental organizations, and educational institutions worldwide. For a partial list of available Papers, see back pages.

DATA from all graphs and tables contained in this Paper are available on 3 1/2" Macintosh or IBM-compatible computer disks. The disks also include data from the *State of the World* series, *Vital Signs*, *Environmental Alert* book series, Worldwatch Papers, and *World Watch* magazine. Each yearly subscription includes a mid-year update, and *Vital Signs* and *State of the World* as they are published. The disk is formatted for Lotus 1-2-3, and can be used with Quattro Pro, Excel, SuperCalc, and many other spreadsheets. To order, see back pages.

© Worldwatch Institute, 1996
Library of Congress Catalog Number 96-060881
ISBN 1-878071-34-3

Printed on 100-percent non-chlorine bleached, partially recycled paper.

Table of Contents

For one-time academic use of this material, please contact Customer Service, Copyright Clearance Center, at (508) 750-8400 (phone), or (508) 750-4744 (fax), or write to CCC, 222 Rosewood Drive, Danvers, MA 01923. Nonacademic users, please call the Worldwatch Institute's Communication Department at (202) 452-1992, x520, or fax your request to (202) 296-7365.

The views expressed are those of the author(s) and do not necessarily represent those of the Worldwatch Institute, its directors, officers, or staff, or of its funding organizations.

ACKNOWLEDGMENTS: I am grateful to water specialists Janusz Kindler, Jan Lundqvist, and Amy Vickers, as well as to Worldwatch colleagues Janet Abramovitz, Lester Brown, Christopher Flavin, Gary Gardner, William Mansfield, and Jim Perry for their valuable comments on an earlier draft of this paper. I also owe thanks to Worldwatch staff members Lori Ann Baldwin and Laura Malinowski for library support; Suzanne Clift, Denise Byers Thomma, Jim Perry, and Jennifer Seher for their assistance with production and outreach; and to Jane Peterson for patience and persistence during editing. Thanks, too, to my many colleagues around the world for continually expanding my knowledge of water issues.

SANDRA POSTEL is Director of The Global Water Policy Project in Amherst, Massachusetts, where her research focuses on international water issues and strategies. For six years, she served as Vice President for Research at Worldwatch Institute, where she remains affiliated as Senior Fellow. She is author of *Last Oasis: Facing Water Scarcity*, selected by *Choice* as one of the outstanding academic books of 1993, and now the basis for a documentary film to air on PBS in 1997. She has published widely in scholarly and popular publications; lectured at Stanford, Harvard, Duke, and Yale universities; and, for the last two years, was Adjunct Professor of International Environmental Policy at Tufts University. She has served on the Board of Directors of the International Water Resources Association and the World Future Society, as an advisor to the Global 2000 program founded by President Jimmy Carter, and as a member of the founding committee of the World Water Council. In 1995, she was awarded a Pew Fellowship in Conservation and the Environment.

Introduction

Deep in the delta of the Colorado River, the Cocopa people have fished and farmed for perhaps 2,000 years. They harvested a grain they called nipa, a unique salt-loving plant known to botanists as *Distichlis palmeri* that tastes much like wild rice. Protein was abundant: they sometimes ate fish three times a day, and they hunted deer, wild boar, ducks, and geese. Known as "people of the river," the Cocopa had no formal calendar, but they keyed their lives to the Colorado's seasonal floods. While no census exists to document their numbers, historical accounts suggest that about 5,000 Cocopa were living in the delta 400 years ago.

Today the Cocopa culture is at risk of extinction. Their millenia-old way of life hangs in the balance because water has been siphoned away to fill swimming pools in Los Angeles, generate electricity to illuminate Las Vegas, and irrigate crops in the deserts of Arizona, California, and the Mexicali Valley. Fishing and farming can no longer sustain them. They last harvested nipa in the early 1950s; by then, dams upstream had largely eliminated the annual floods that had naturally irrigated their staple grain. Now, just 40-50 Cocopa families remain south of the border. With little means of subsistence or livelihood in the delta countryside, many of the tribal members have migrated to the cities. Anita Alvarez de Williams, a Mexicali-based expert on the Cocopa, worries that by the end of the 20th century they "may no longer be river people at all."[1]

It might be tempting to dismiss the Cocopa's plight as

a price of progress. Supporting ever higher levels of consumption and population has always involved taking more and more of nature's water bounty, and those last in line are bound to suffer. But apart from the tragedy of losing yet another culture in a world of dwindling cultural diversity, the fading of the Cocopa is a harbinger—an early warning of larger-scale disruption.

A growing scarcity of fresh water is now a major impediment to food production, ecosystem health, social stability, and peace among nations. Each year, millions of tons of grain are grown by depleting underground aquifers—a clear case of robbing the future to pay for the present. In Asia, where most of the world's population growth, and thus additional food needs, will occur, many rivers run dry for all or part of the dry season—including the mighty Ganges, a key water source for the subcontinent's rapidly growing population. In 1995, the lower stretch of China's Yellow River was parched for a third of the year. Africa now has nearly 300 million people living in water-stressed countries, many of which are already heavily dependent on food imports. Under current population projections, the number of Africans living in water-stressed countries will grow to more than 1.1 billion by 2025, three-quarters of the continent's projected population at that time.

As supplies dwindle, competition for water is increasing—between cities and farms, between neighboring states and provinces, and between nations. Tensions over water persist in the major river basins of the Middle East—the Nile, the Jordan, and the Tigris-Euphrates—as well as in the Ganges of South Asia and the Aral Sea basin of central Asia. In none of these is there yet a treaty recognized by all parties that allocates the basin's waters among them. Worldwide, at least 214 rivers flow through two or more countries, yet no enforceable law governs the allocation and use of international waters.

Rivers, lakes, and wetlands are in declining health because the traditional approach to water development has failed to protect their vital ecological functions—including flood protection, water purification, habitat maintenance, and sustenance of fisheries. In the United States, 37 percent of freshwater fish species are at risk, primarily because of the destruction and degradation of their habitats. Along with the Colorado River Delta, the Nile, Ganges, Amu Dar'ya and Syr Dar'ya deltas are among those experiencing serious deterioration from large-scale damming and diverting of their rivers. The Aral Sea has shrunk in volume by three-fourths as a result of excessive diversions from the Amu Dar'ya and the Syr Dar'ya. It has lost 20 of its 24 fish species, along with a fish catch that once totaled 44,000 tons a year and supported 60,000 jobs.

Although fresh water is renewable, it is also finite.

For all its impressive engineering, modern water development has adhered to a fairly simple formula: estimate the demand for water and then build new supply projects to meet it. It is an approach that ignores concerns about human equity, the health of ecosystems, other species, and the welfare of future generations. In a world of resource abundance, it may have served humanity adequately. But in the new world of scarcity, it is fueling conflict and degradation.

As world population expands by a projected 2.6 billion people over the next 30 years, and as consumption levels spiral upward, water problems are bound to intensify. Three principle challenges stand out: securing a sufficient and sustainable supply of water for food production, arresting the decline of aquatic ecosystems, and averting social unrest and political conflict as competition for water increases. With the best dam sites already developed and many groundwater reserves overtapped, opportunities to meet these challenges by exploiting new water sources are clearly limited. A new and more promising blueprint is needed, one with goals of satisfying basic human and ecological needs, using and allocating the remaining water more efficiently, and sharing international waters equitably.[2]

At the moment, the policies, institutions, and planning

procedures in place to manage water are not well suited to these tasks. A flurry of international activity in recent years has called attention to water scarcity and the need for action to address it. Agenda 21, the global action plan that emerged from the 1992 Earth Summit in Rio de Janeiro, and the World Bank's 1993 water resources policy paper put forth helpful guiding principles and spawned a variety of national studies. The United Nations Commission on Sustainable Development has requested a global freshwater assessment, which is now underway and is due to be reported to the U.N. General Assembly in 1997. But most nations have not realistically assessed how limited water supplies will affect their food production and economic prospects. Priorities need to be set, and tradeoffs made. In short, the challenging task of charting a sustainable course for water use and management remains.[3]

More rational water pricing, carefully monitored water markets, the setting of efficiency standards, and the allocation of water to meet minimum requirements for human and ecosystem health are some of the policy tools and actions that are needed. Meanwhile, a host of conservation and efficiency measures with the potential to save vast quantities of water remain untapped because of inadequate incentives to encourage their use. The newly formed World Water Council is taking on the concrete task of establishing water-use efficiency targets for various activities, a potentially useful step forward. Also at the international level, new impetus is needed for countries in water-scarce river basins to negotiate water-sharing agreements and to establish mechanisms for joint river basin management. Only with such arrangements can tensions be resolved and the benefits of more optimal management be realized.[4]

Policymakers have vastly underestimated the influence of water scarcity on economic progress, food security, and regional peace and stability. Many have yet to realize that water problems can no longer be fixed by engineers alone. The challenges water scarcity poses are complex. They require that political leaders take notice, and act.

Global Mirage

The photographs of earth taken by astronauts show a strikingly blue planet, seemingly a water world spinning in space. As it turns out, however, this impression of water wealth may be as deceptive as a desert mirage. Only about 2.5 percent of all the water on earth is fresh, and two-thirds of this fresh water is locked in glaciers and ice caps. The annual renewable freshwater supply on land—that made available year after year by the solar-powered hydrologic cycle in the form of precipitation—totals some 110,000 cubic kilometers, a mere 0.000008 percent of all the water on earth. (A cubic kilometer equals a billion cubic meters.)[5]

Each year, nearly two-thirds of this renewable supply returns to the atmosphere through evaporation or transpiration, the uptake and release of moisture by plants. This evapotranspiration (it is difficult to distinguish the two processes over large areas; hence, the joint term) represents the water supply for forests, grasslands, rain-fed croplands, and all other non-irrigated vegetation. Just over one-third of the renewable supply—about 40,000 cubic kilometers per year—is runoff, the flow of fresh water from land to sea through rivers, streams, and underground aquifers. Runoff is the source for all human diversions or withdrawals of water—what is typically called water "demand" or water "use." It is the supply for irrigated agriculture, industry, and households, as well as for a wide variety of "instream" water services, including the maintenance of freshwater fisheries, navigation, the dilution of pollutants, and the generation of hydroelectric power. Rivers also carry nutrients from the land to the seas, and in this way help support the highly productive fisheries of coastal bays and estuaries. Thus, by virtue of the hydrologic cycle, the oceans water the continents, and the continents nourish the oceans.[6]

Although fresh water is renewable, it is also finite: the land receives roughly the same amount of water today as when the first civilizations emerged thousands of years ago.

Moreover, nature's delivery of water does not match up well
with the distribution of world population. Asia, for example, has 36 percent of global runoff but 60 percent of the
world's people; South America, on the other hand, supports
6 percent of the world's people and has 26 percent of the
world's runoff. (See Table 1.) In addition, much of the river
flow in the tropics and high latitudes is virtually inaccessi-
ble to people and economic activity and is likely to remain
so for the foreseeable future because water is difficult and
expensive to transport long distances. The Amazon River
alone carries 15 percent of the earth's runoff but is accessible to only 0.4 percent of world population. Fifty-five rivers
in northern North America, Europe, and Asia, with combined annual flows equal to about 5 percent of global
runoff, are so remote that they have no dams on them at
all.[7]

River flows are also highly uneven over time and are
often poorly matched to the pattern of human need for
water. Water is only useful to farmers, industries, and cities
if it can be tapped and used more or less on demand. Since
about three-fourths of runoff is floodwater, engineers have
built dams in an attempt to capture and store a portion of it
for later use. Today, large dams, which collectively can hold
about 14 percent of annual runoff, have increased the stable
supply of water provided by underground aquifers and year-
round river flows by nearly a third, bringing the total stable
renewable supply to 14,600 cubic kilometers. Of this total,
an estimated 12,500 cubic kilometers is within reach geo-
graphically, and this is the amount currently accessible for
irrigation, industries, and households.[8]

Globally, water use roughly tripled between 1950 and
1990, and now stands at an estimated 4,430 cubic kilometers—35 percent of the accessible supply. At least an additional 20 percent is used "instream" to dilute pollution, sustain fisheries, and transport goods. So humanity is already
using, directly or indirectly, more than half of the water supply that is now accessible.[9]

Given that world population is projected to climb by

TABLE 1

Global Runoff and Population, by Continent, 1995

Region	Total Annual Runoff (cubic kilometers)	Share of Global Runoff (percent)	Share of Global Population (percent)
Europe	3,240	8	13
Asia	14,550	36	60
Africa	4,320	11	13
N. and C. America	6,200	15	8
S. America	10,420	26	6
Australia & Oceania	1,970	5	<1
Total	40,700	101[1]	~100[1]

[1]Does not add to 100 because of rounding.
Source: See endnote 7.

2.6 billion over the next 30 years, about the same number that was added between 1950 and 1990, this is a troubling finding. Worldwide water use cannot triple again without a substantial increase in the available supply, much greater reuse of the existing supply, and near-universal efforts to prevent pollution. However, the construction of new dams has slowed markedly over the last couple of decades as the public, governments, and financial backers have begun to pay more attention to their high economic, social, and environmental costs. Whereas nearly 1,000 large dams came into operation each year from the 1950s through the mid-1970s, the number dropped to about 260 annually during the early 1990s. Even if conditions become more favorable to dam construction, it seems unlikely that new reservoirs built over the next 30 years will increase accessible runoff by more than 10 percent.[10]

With the oceans holding more than 97 percent of all the water on earth, desalination is often held up as the ultimate solution to the world's water problems. Indeed, as early as 1961, U.S. President John F. Kennedy noted that if humanity could find an inexpensive way to get fresh water from the seas, it would be an achievement that "would really dwarf any other scientific accomplishment."[11]

Thirty-five years later, desalination is a proven technology experiencing solid growth. As of December 1995, a total of 11,066 desalting units had been installed or contracted for worldwide, with a collective capacity of 7.4 billion cubic meters per year. Total capacity expanded at an average annual rate of about 4.5 percent over the previous two years, compared with an average rate of 10 percent per year from December 1989 through December 1993. Analysts attribute the recent slowdown to budget constraints in Middle Eastern countries, which prevented them from ordering any large desalination plants.[12]

Despite considerable growth, desalination still plays a minor part in the global supply picture—accounting for less than two-tenths of 1 percent of world water use. Removing salt from water is highly energy intensive, and, although costs have come down, at $1.00-$1.60 per cubic meter, desalination remains one of the most expensive supply options. Saudi Arabia, United Arab Emirates, and Kuwait—which together have only 0.4 percent of world population—accounted for 46 percent of the world's 1993 desalting capacity. In a sense, these countries are turning oil into water, and they are among the few that can afford to do so. For the foreseeable future, seawater desalination will likely continue to be a lifeline technology for water-scarce, energy-rich countries, as well as island nations with no other options, but a minor contributor to total water supplies worldwide.[13]

Even this quick assessment shows that the widespread impression that water is too abundant globally to constrain human activities is ill founded. Before long, the consequences of water scarcity will spread beyond the specific regions, such as the Middle East and much of Africa, that are bumping into water's limits today. As world population expands by a projected 45 percent over the next 30 years, there will be global effects—on food production, ecological support systems, social stability, and geopolitics—that will reverberate throughout the world economy.

Water for Food

Water is the basis of life—one of those facts so fundamental it is easy to forget. Using the sun's energy, plants combine water with carbon dioxide to form carbohydrates, the earth's basic food supply. Without water, life and growth cease—a stark reality that may be taking on ever greater importance: just as the world's food needs are climbing by record quantities, the amount of fresh water that can sustainably be supplied to farmers is nearing its limits.

Crop production is a highly water-intensive activity. Worldwide, agriculture accounts for about 65 percent of all the water removed from rivers, lakes, and aquifers for human activities, compared with 22 percent for industries, and 7 percent for households and municipalities. (See Table 2.) Moreover, whereas homes and factories return a large portion of their water to the environment after they use it (albeit often in a polluted state), half to two-thirds of agriculture's share is "consumed" through evaporation or transpiration, and is thus not available for a second or third use.[14]

Producing a ton of harvested grain consumes about 1,000 tons of water. The actual amount will vary with the type of grain and the climate in which it is grown, but this is a reasonable average. This figure includes the crops' evapotranspiration needs, but not water lost because of inefficiencies in irrigation methods. As such, it represents an approximate minimum water requirement for the production of grain, the source of roughly half of the calories humans consume directly, and two-thirds of all calories they consume, including those derived from grain-fed livestock products.[15]

As of 1995, annual global grain consumption averaged

TABLE 2

Estimated Global Water Demand and Consumption, by Sector, c. 1990

Sector	Estimated Demand	Share of Total	Estimated Consumption	Share of Total
	(cubic kilometers per year)	(percent)	(cubic kilometers per year)	(percent)
Agriculture[1]	2,880	65	1,870	82
Industry[2]	975	22	90	4
Municipalities[2]	300	7	50	2
Reservoir Losses[3]	275	6	275	12
Total	4,430	100	2,285	100

[1]Assumes average applied water use of 12,000 cubic meters per hectare of irrigated land, and consumption equal to 65 percent of demand. [2]Estimates are from Shiklomanov; see endnote 5. [3]Assumes evaporation loss equal to 5 percent of gross reservoir storage capacity.
Source: Sandra L. Postel, Gretchen C. Daily, Paul R. Ehrlich, "Human Appropriation of Renewable Freshwater," *Science,* February 9, 1996.

about 300 kilograms per person. Assuming, conservatively, that the global average remains at today's level, just meeting the grain requirement of the projected world population in 2025 would take an additional 780 billion cubic meters of water.[16]

Estimating the amount of water required to grow the other crops in the human diet is difficult, both because diets vary and because the water-use efficiency of crops varies. In general, fruits and vegetables require less water per ton of harvested yield (because they are mostly water) than more-nutritious rice, wheat, or corn do, and crops grown in cooler climates or seasons require less than ones grown in hotter places or times. Assuming that the non-grain portion of the diet requires a third as much water to produce as the grain-based portion does, then the minimum amount of water needed to produce an average diet would be about 400 cubic meters per person per year. At this level, meeting the food requirements of the 2.6 billion people expected to be added to the planet by 2025 would take an additional 1,040 billion cubic meters of water—equal to more than 12 times the average flow of the Nile River, or 56 times the average flow of the Colorado River.[17]

Where this water is to come from on a sustainable basis is not obvious. Crops get the moisture they need from natural rainfall, irrigation, or some combination of these two sources. Because the potential for expanding cropland area is limited, the bulk of additional food needs will have to be met through increases in crop yields, as has been the case since mid-century: yield increases accounted for nearly 80 percent of increased output between 1950 and 1981, the year that world grainland area peaked. Along with the "green revolution" technologies of high-yielding crop varieties, fertilizers, and pesticides, the spread of irrigation contributed greatly to the surge in yields. Irrigation enables farmers to apply water to their fields at desired times and in desired amounts, a degree of control needed for intensive, high-yielding production. It also typically makes possible two or three harvests per year from the same parcel of land. As a result, irrigated lands are disproportionately important to global food security: they account for only 16 percent of the world's cropland, but they yield some 40 percent of the world's food.[18]

There is now abundant evidence, however, that the heyday of irrigation expansion is over. Worsening imbalances between population levels and available water supplies, falling water tables, depleted river flows, the lack of economical and environmentally sound sites for new supply projects, and rapidly growing urban demands are all constraining the amount of water available for agriculture.

One measure of water constraints is derived by comparing a country's renewable water supply with its population size. As a rule of thumb, countries are considered "water stressed" when their renewable runoff per person drops below 1,700 cubic meters per year; they are considered "water scarce" when it drops below 1,000 cubic meters per year. Exactly what "stressed" and "scarce" mean is not clear, and water analysts heatedly debate the terms. Hillel Shuval of Hebrew University points out, for example, that Israel has a highly successful modern economy and high per capita income even though its renewable water per person is less than a fifth of the water-stress level. In part, Israel has succeeded so well with its limited supplies by importing much of its grain—which Shuval and others sometimes refer to as "virtual water."[19]

Indeed, with each ton of grain representing 1,000 tons of water, importing grain becomes a key strategy for balancing water budgets. Such a strategy would seem to make economic and environmental sense for countries short of water, since they can get much higher value from their limited supplies by devoting them to commercial and industrial enterprises, and then using the resulting income to purchase food through international markets. The Middle East, for instance, which is the most concentrated region of water scarcity in the world, imports 30 percent of its grain. As long as surplus food is produced elsewhere, nations with surpluses are willing to trade, and the countries in need can afford to pay for the imports, it would seem that water-short countries can have food security without needing to be food self-sufficient.[20]

This tidy logic is shaken, however, by the growing number of people living in countries where water availability is a constraint to food self-sufficiency, and by widespread signs of unsustainable water use in key food-producing regions. At runoff levels below 1,700 cubic meters per person per year, food self-sufficiency is likely to be problematic, if not impossible. In many, if not most, countries it is difficult to access and control more than 30 percent of runoff. (There are, however, notable exceptions, such as Egypt, with the massive storage provided by the Aswan Dam.) Per capita runoff of 1,700 cubic meters annually would thus translate to 510 cubic meters of usable supply to meet irrigation, industrial, and household needs. If 125 cubic meters per person per year is needed for domestic, urban, commercial, and industrial needs, and 400 cubic meters is required to support an average diet, then a minimum of 525 cubic meters is needed to support each person for a year—just above the level likely to be accessible in a country with 1,700 cubic meters per person of runoff.[21]

As of 1995, a total of 44 countries with a combined population of 733 million people had annual renewable water supplies per person below 1,700 cubic meters. Just over half of these people live in Africa or the Middle East, where the populations of many countries are projected to double within 30 years. (See Table 3.) Water-short Algeria, Egypt, Libya, Morocco, and Tunisia are each already importing more than a third of their grain. With their collective population projected to grow by 87 million people over the next 30 years, their dependence on grain imports is bound to increase. Indeed, this is a likely scenario for much of Africa: given current population projections, more than 1.1 billion Africans will be living in water-stressed countries by 2025—three-quarters of the continent's projected population at that time.[22]

Water-short Algeria, Egypt, Libya, Morocco, and Tunisia are already importing more than a third of their grain.

Parts of many large countries, including China, India, and the United States, would qualify as water stressed if breakdowns of water supplies and population were available by region. Even using national statistics, China—with 7 percent of global runoff but 21 percent of world population—will only narrowly miss the 1,700 cubic meter per capita mark in 2030; India, the world's second most populous country, will join the list by then.[23]

Many physical signs of unsustainable water use authenticate this theoretical indicator of water stress, and provide complementary evidence of limits to agricultural water use as well. Groundwater overpumping and aquifer depletion are now occurring in many of the world's most important crop-producing regions—including the western

TABLE 3

African and Middle Eastern Countries with Less than 1,700 Cubic Meters of Runoff per Person, 1995, with Population Projections to 2025[1]

Country	Internal Runoff Per Capita, 1995 (cubic meters per year)	1995 Population (millions)	Projected 2025 Population (millions)
AFRICA			
Algeria	489	28.4	47.2
Burkina Faso	1,683	10.4	20.9
Burundi	563	6.4	13.5
Cape Verde	750	0.4	0.7
Djibouti	500	0.6	1.1
Egypt	29	61.9	97.9
Eritrea	800	3.5	7.0
Kenya	714	28.3	63.6
Libya	115	5.2	14.4
Mauritania	174	2.3	4.4
Morocco	1,027	29.2	47.4
Niger	380	9.2	22.4
Rwanda	808	7.8	12.8
Somalia	645	9.3	21.3
South Africa	1,030	43.5	70.1
Sudan	1,246	28.1	58.4
Tunisia	393	8.9	13.3
Zimbabwe	1,248	11.3	19.6
MIDDLE EAST			
Iraq	1,650	20.6	52.6
Israel	309	5.5	8.0
Jordan	249	4.1	8.3
Kuwait	0	1.5	3.6
Lebanon	1,297	3.7	6.1
Oman	909	2.2	6.0
Saudi Arabia	119	18.5	48.2
Syria	517	14.7	33.5
United Arab Emirates	158	1.9	3.0
Yemen	189	13.2	34.5
Total 1995		380.6	739.8
Additional Countries By 2025[2]			618.0
Projected Total 2025			1,357.8

[1] Runoff figures do not include river inflow from other countries: Djibouti, Mauritania, Sudan, and Iraq would have more than 1,700 cubic meters per person in 1995 and 2025 if current inflow from other countries were included.
[2] Botswana, Ethiopia, Gambia, Ghana, Lesotho, Madagascar, Malawi, Mauritius, Nigeria, Senegal, Swaziland, Tanzania, Togo, Uganda.
Source: See endnote 22.

United States and large portions of India, as well as parts of north China, where water tables are dropping 1 meter a year over a large area. (See Table 4.) Not only does this signal that limits to groundwater use have been exceeded in many areas, it means that a portion of the world's current food supply is produced by using water unsustainably—and can therefore not be counted as reliable over the long term. In the Punjab, India's breadbasket, for example, groundwater tables are dropping by 20 centimeters annually over two-thirds of the state, and according to researchers at Punjab Agricultural University, "'questions are now being asked as to what extent rice cultivation should be permitted in the irrigated Indo-Gangetic Plains, and how to sustain the productivity of the region without losing the battle on the water front.'"[24]

In some cases, groundwater depletion permanently reduces the earth's natural capacity to store water. The extraction of water may cause an aquifer's geologic materials to compact, eliminating the pores and spaces that held the water. The loss of this storage capacity is irreversible, and it carries a high cost. In California, for example, compaction of overdrafted aquifers in the Central Valley has resulted in a loss of nearly 25 billion cubic meters of storage capacity— equal to more than 40 percent of the combined storage capacity of all human-made surface reservoirs statewide. The economic value of this loss is difficult to calculate; however, assuming a replacement cost of 24¢ per cubic meter, which is at the low end of the range of costs for new water storage options in California, the economic value of the destroyed groundwater storage capacity would amount to $6 billion.[25]

Like groundwater, many of the planet's major rivers are suffering from overexploitation. In Asia, where the majority of population growth and additional food needs will be centered, many rivers are completely tapped out during the drier part of the year, when irrigation is so essential. According to a 1993 World Bank study, "many examples of basins exist throughout the Asia region where essentially no

TABLE 4

Groundwater Depletion in Major Regions of the World, c. 1990

Region/Aquifer	Estimates of Depletion
High Plains Aquifer System, United States	Net depletion to date of this large aquifer, which underlies nearly 20% of all U.S. irrigated land, totals some 325 billion cubic meters, roughly 15 times the average annual flow of the Colorado River. More than two-thirds of this depletion has occurred in the Texas High Plains, where irrigated area dropped by 26% between 1979 and 1989. Current depletion is estimated at 12 billion cubic meters per year.
California, United States	Groundwater overdraft averages 1.6 billion cubic meters per year, amounting to 15% of the state's annual net groundwater use. Two-thirds of the depletion occurs in the Central Valley, the country's (and to some extent the world's) fruit and vegetable basket.
Southwest United States	Overpumping in Arizona alone totals more than 1.2 billion cubic meters per year. East of Phoenix, water tables have dropped more than 120 meters. Projections for Albuquerque, N.M., show that if groundwater withdrawals continue at current rates, water tables will drop an additional 20 meters on average by 2020.
Mexico City and Valley of Mexico	Pumping exceeds natural recharge by 50-80%, which has led to falling water tables, aquifer compaction, land subsidence, and damage to surface structures.
Arabian Peninsula	Groundwater use is nearly three times greater than recharge. Saudi Arabia depends on nonrenewable groundwater for roughly 75% of its water, which includes irrigation of 2-4 million tons of wheat per year. At the depletion rates projected for the 1990s, exploitable groundwater reserves would be exhausted within about 50 years.
North Africa	Net depletion in Libya totals nearly 3.8 billion cubic meters per year. For the whole of North Africa, current depletion is estimated at 10 billion cubic meters per year.
Israel and Gaza	Pumping from the coastal plain aquifer bordering the Mediterranean Sea exceeds recharge by some 60%; salt water has invaded the aquifer.

Spain	One-fifth of total groundwater use, or 1 billion cubic meters per year, is unsustainable.
India	Water tables in the Punjab, India's breadbasket, are falling 20 centimeters annually across two-thirds of the state. In Gujarat, groundwater levels declined in 90% of observation wells monitored during the 1980s. Large drops have also occurred in Tamil Nadu.
North China	The water table beneath portions of Beijing has dropped 37 meters over the last four decades. Overdrafting is widespread in the north China plain, an important grain-producing region.
Southeast Asia	Significant overdraft has occurred in and around Bangkok, Manila, and Jakarta. Overpumping has caused land to subside beneath Bangkok at a rate of 5-10 centimeters a year for the past two decades.

Source: Global Water Policy Project and Worldwatch Institute, based on sources in endnote 24.

water is lost to the sea during much of the dry season." These include most rivers in India—among them, the Ganges, a principal water source for densely populated and rapidly growing South Asia. China's Yellow River has gone dry in its lower reaches for an average of 70 days a year in each of the last ten years, and in 1995, the river was dry for 122 days. Demand for water is exceeding the river's capacity to supply it, and crop production in the region is bound to suffer increasingly.[26]

Another indication that agricultural water supplies are limited comes from examining global irrigation trends. Rising water development costs and the declining number of environmentally sound sites for the construction of dams and river diversions are contributing to a worldwide slowdown in irrigation expansion. Per capita irrigated area was steady or increasing during most of modern times, but it peaked in 1979 and has declined by about 7 percent since then. (See Figure 1.) This irrigation slowdown, which partly explains the decline in per capita grain production that has occurred since 1984, is a worrisome trend that is not likely to reverse anytime soon.[27]

At the same time, much irrigated land is losing productivity or coming out of production altogether as a result of salinization, the steady buildup of salts in the root zone of irrigated soils. Although no firm global estimate exists, some 25 million hectares—more than 10 percent of world irrigated area—likely suffer from salt buildup serious enough to lower crop yields. Salinization is estimated to be spreading at a rate of up to 2 million hectares a year, offsetting a good portion of the gains achieved by irrigation expansion. Indeed, David Seckler, Director General of the International Irrigation Management Institute in Sri Lanka, writes in a 1996 paper that "the net growth of irrigated area in the world has probably become *negative.*"[28]

Finally, agriculture is losing some of its existing water supplies to cities as population growth and urbanization push up urban water demands. Worldwide, the number of urban dwellers is expected to double, to 5 billion, by 2025. With political power and money concentrated in the cities, and with insufficient water to meet all demands, governments will face strong pressures to shift water out of agriculture—even as food demands are rising. Where this shift results from marginal lands or nonfood crops coming out of production, or from gains in irrigation efficiency, it can be beneficial environmentally and have little impact on food security. But with competition for water increasing in many areas, sizable shifts could reduce food production and destabilize farm communities.[29]

The reallocation of water from farms to cities is well underway in both industrial and developing countries. In California, for instance, a 1957 state water plan projected that 8 million hectares of irrigated land would ultimately be developed statewide; yet the state's irrigated area peaked in 1981 at 3.9 million hectares, less than half this amount. Net irrigated area fell by more than 121,000 hectares during the 1980s. California officials project an additional net decline of nearly 162,000 hectares between 1990 and 2020, with

World Irrigated Area per Thousand People, 1961–93

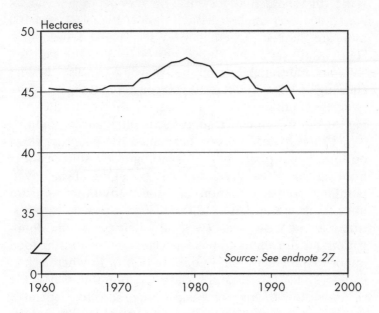

Source: See endnote 27.

most of the loss due to urbanization as the population expands from 30 million to a projected 49 million.[30]

In China, water supplies are being siphoned away from farmlands surrounding Beijing in order to meet that city's rising urban, industrial, and tourist demands. The capital's water use now exceeds the capacity of its two main reservoirs, and so farmers in the agricultural belt that rings the city have been cut off from their traditional sources of irrigation water. With some 300 Chinese cities now experiencing water shortages, this shift is bound to become more pronounced.[31]

Growing demand in the megacities of Southeast Asia—including Bangkok, Manila, and Jakarta—is already partially being met by overpumping groundwater. With limited new sources to tap, pressures to shift water out of agriculture will mount in these regions as well. These and other rapidly industrializing regions also face the double-edged sword of rising affluence and the higher consumption levels it engenders, which will intensify competition for agriculture's water. In Malaysia, for instance, the number of golf courses has tripled over the last decade to more than 150, and 100 more are planned. Together, Malaysia, Thailand, Indonesia, South Korea, and the Philippines boast 550 golf courses, with an additional 530 on the drawing boards. Besides chewing up farms and forests, golf courses in these countries typically require irrigation at the same time crops do—during the dry season, when supplies are often already tight.[32]

Unfortunately, no one has tallied the potential effect on future food production of the progressive shift of water from agriculture to cities, combined with groundwater over-pumping, aquifer depletion, and the many other forms of unsustainable water use. Without such assessments, countries have no clear idea how secure their agricultural foundations are, no ability to predict accurately their future food import requirements, and no sense of how or when to prepare for the economic and social disruption that may ensue as farmers lose their water. Klaus Lampe, Director General of the International Rice Research Institute (IRRI) in the Philippines, warns: "Thoughtlessness and ignorance regarding tomorrow's food supply are among the most dangerous of the many factors influencing our political, economic, and environmental systems."[33]

As discussed in the last section of this paper, broad policy reforms will be needed to encourage more efficient and productive use of water in agriculture so as to make better use of the limited supplies available; water and agricultural scientists cannot solve these problems alone. Scientific research, however, is critical—especially in raising the water productivity of the global crop base. If the battle on the agricultural water front is to be won, crop output per unit of water input will need to increase not only in irrigated farming systems, but in rainfed and water-harvesting systems as well. The actual strategies used will vary by crop, climate, and the type of water-control system, but the basic aim will necessarily be the same in each: to optimize the timing and amount of moisture in the root zone and to enhance the crops' ability to use that moisture productively.

Through plant breeding and genetic manipulation, for example, scientists can hasten the process of plant adaptation to dryness. Studies have shown that, if no other factors are limiting, total dry-matter production is linearly proportional to the amount of water a plant evapotranspires. Larger or deeper root systems that allow plants to take in more moisture can thus increase yield; so can adapting crops for early reproduction so as to avoid late-season droughts. New genetic techniques are making it possible to screen crop varieties for water-efficiency traits. And developing varieties with shorter growing seasons or the ability to grow in cooler periods, when evapotranspiration is lower, could also help improve crop water use efficiency.[34]

The International Rice Research Institute and the International Irrigation Management Institute in Sri Lanka are two of the research centers taking up the water-efficiency challenge in a major way. A 1995 IRRI report notes that "while the full import of the water supply problem to rice production has been recognized only relatively recently by the research community, it is now fully acknowledged there." Future rice production will depend heavily on getting "more rice per unit of water input." IRRI is focusing on developing more efficient irrigation operations, technologies that reduce water consumption, and changes in the rice plant itself to improve water-use efficiency. Breeders have already shortened the maturation time for irrigated rice from 150 days to 110 days, for example, a major water-saving achievement.[35]

Better matching crops to varying qualities of water can enhance water supplies for irrigated agriculture. In the western Negev of Israel, for example, cotton is successfully grown by irrigation with highly salty water from a local saline aquifer. The Israelis have also found that certain crops—such as tomatoes grown for canning or pastes—may actually benefit from somewhat salty irrigation water. The various salt tolerances of crops raise the possibility of multiple reuse of irrigation water. In California, for instance, moderately salty drainage water from a crop of average salt tolerance is used to irrigate more highly tolerant cotton. In turn, the drainage from the cotton fields, which is highly salty, is used to irrigate a halophytic (salt-loving) crop, a number of which scientists have made considerable progress toward commercializing. A variety of Salicornia, for example, was irrigated with seawater in a coastal desert near Mexico's Sea of Cortez (also known as the Gulf of California) and yielded seed and biomass equal to or greater than fresh-water oilseed crops such as soybean and sunflower.[36]

Water for Ecosystems

A low-altitude flight over the Colorado River Delta, not far from where the Cocopa traditionally harvested their nipa, reveals the dry channel of the Colorado, which still traces the river's meandering path toward the sea. This is the place where, for millennia, the river deposited its rich load of silt and supported a diverse ecosystem before delivering its treasure of nutrients to the upper Sea of Cortez. It is the place where American naturalist Aldo Leopold journeyed by canoe in 1922 and reported seeing deer, quail, raccoon, bobcat, and vast fleets of waterfowl. The winding river, slowing as it spread out through countless green lagoons, later led Leopold to muse, "for the last word in procrastination, go travel with a river reluctant to lose his freedom in the sea."[37]

Leopold never returned to the delta for fear of finding this "milk-and-honey wilderness" badly altered—and his fears were justified. Today, the Colorado's freedom has been lost to a degree even the prescient Leopold could scarcely have imagined. Only in years of extremely high precipitation in its watershed does the Colorado run all the way to the sea. In most years, what remains of its flow after ten major dams and several large diversions is a trickle that lit-

FIGURE 2

Flow of Colorado River below All Major Dams and Diversions, 1905–1992

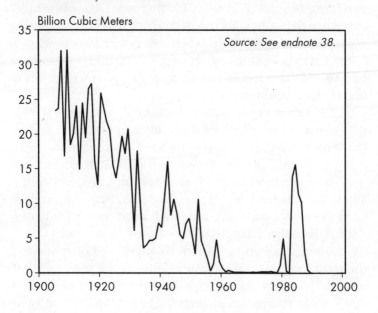

Billion Cubic Meters

Source: See endnote 38.

erally disappears into the desert of northern Mexico. (See Figure 2.) Much of the delta's once-abundant wildlife is gone. Fisheries in the Sea of Cortez have declined dramatically. And the native Cocopa way of life is threatened with extinction.[38]

What has happened to the Colorado is but one example of a disturbing and widespread decline of the aquatic environment. Globally, the tripling of water use since 1950 has led to the building of more and bigger water supply projects—especially dams and river diversions. Around the world, the number of large dams (those more than 15 meters high) built for water supply, hydropower, and flood control has climbed from just over 5,000 in 1950 to roughly 38,000 today. More than 85 percent of the large dams now standing have been built during the last 35 years.[39]

This is a massive change in the global aquatic environment in a very short period of time. Many rivers now resemble elaborate plumbing works, with the timing and amount of flow completely controlled, like water from a faucet, to maximize the rivers' benefits for humans. But while modern engineering has been remarkably successful at getting water to people and farms when and where they need it, it has failed to protect the fundamental ecological functions of rivers and freshwater ecosystems—many of which go unvalued in the marketplace and are easy to take for granted.

The age-old notion that any runoff to the sea is "wasted" remains a dominant view among many water engineers, but it reflects a narrow sense of what a river's role and functions really are. Rivers deliver nutrients to the seas, with their complex food webs; sustain economically and culturally important fisheries; protect wetlands, with their capacity to filter out pollutants; provide habitat for a rich diversity of aquatic life; safeguard fertile deltas; protect water quality; maintain salt and sediment balances; and offer some of the most inspirational natural beauty on the planet. These benefits and services are rapidly being lost because water development plans and projects fail to account for and value them.[40]

In the United States, the most comprehensive assessment to date of the status of native plant and animal species notes that the most striking finding is "the dire condition of those species that depend on aquatic systems for all or part of their life cycle." Out of 13 groups, the four with the greatest share of species at risk are freshwater mussels (67 percent at risk), crayfish (65 percent), amphibians (38 percent), and freshwater fishes (37 percent). For comparison, the study found that the share of bird, mammal, and reptile species at risk is 14 percent, 16 percent, and 18 percent, respectively. The destruction and degradation of habitats is the leading cause of imperilment.[41]

In California, for instance, one of the most heavily plumbed places on the planet, water development has devastated aquatic systems and the life dependent on them. The state has lost 95 percent of its wetlands, and populations of migratory birds and waterfowl, which depend on such areas for food and habitat, have dropped from 60 million around 1950 to 3 million today. Extensive river damming and destruction of spawning habitat has greatly reduced fish populations. California's salmon and steelhead population, for example, has fallen by an estimated 80 percent.[42]

The mighty Ganges in South Asia is among the major rivers that no longer reach the sea for all or part of the year. (See Table 5.) India's heavy diversions upstream during the dry season leave almost nothing in the river for Bangladesh, much less enough to reach the river's natural outlet in the Bay of Bengal. The lack of fresh water flowing out to sea has caused the rapid advance of a saline front across the western portion of the river delta, which is damaging valuable mangroves and fish habitat, important resources for local inhabitants. Unless more water is allowed to flow into the delta during the dry season, damage to vegetation and fisheries will continue, spreading disruption to the local economy.[43]

Around the world, the number of large dams has climbed from just over 5,000 in 1950 to roughly 38,000 today.

In the Nile River basin, the High Dam at Aswan was constructed during the 1960s to provide virtually complete control over the Nile's waters and a crucial hedge against drought. Lake Nassar is able to store fully two years' worth of the Nile's average annual flow. Not surprisingly, however, the High Dam has greatly altered the river system. Out of 47 commercial fish species in the Nile prior to the dam's construction, only 17 were still harvested a decade after its completion. The annual sardine harvest in the eastern Mediterranean dropped by 83 percent, likely a side-effect of the reduction in nutrient-rich silt entering that part of the sea.[44]

One of the most worrisome long-term consequences of the disruption of the Nile ecosystem is that the river delta,

TABLE 5

Status of Selected Major Rivers That Are Heavily Dammed and Diverted

River and Source	Annual Flow at Full Strength	Location of Mouth and Condition There	Consequences of Diminished Flow
	(billion cubic meters per year)		
Ganges; Himalaya of Nepal	587	Bangladesh and Bay of Bengal. Little or no flow during dry season.	Rising salt level in delta damages mangroves and fisheries; low flow into Bangladesh is harming crop production; tensions persist between India and Bangladesh.
Nile; Ethiopian Highlands (Blue Nile) and Lake Victoria region (White Nile)	84	Egypt and Mediterranean Sea. Only about 2% of Nile's freshwater flow reaches the sea— all of it during the winter months; remainder of outflow is polluted farm drainage.	Egypt's water demand is soon to outstrip its supply; Nile Delta is subsiding from trapping of silt behind Aswan Dam; no water-sharing agreement exists that includes all ten basin countries.
Amu Dar'ya and Syr Dar'ya; Mountains of Central Asia	69	Aral Sea. Combined annual flow into sea has dropped dramatically because of large diversions for cotton irrigation.	Aral Sea has lost half its area and three-fourths of its volume since 1960; salinity has tripled; 44,000-ton annual fish catch has dropped to zero; disease from contaminated water is rampant; 20 of 24 fish species have disappeared.
Huang He (Yellow River); Tibetan Plateau	56	Bo Hai/Yellow Sea. Lower reaches often dry; in 1995, river was dry for a third of the year.	Competition for water is increasing in the north China plain; the government is planning a costly south-to-north diversion from the Yangtze River.

| Chao Phraya; Thailand | 22 | Gulf of Thailand. With demand exceeding supply, there are potentially low flows during the dry season. | A salt front threatens to advance across the delta; competition for water is intensifying between farm and urban uses; supplies to Bangkok are insufficient to alleviate groundwater overpumping. |
| Colorado; Rocky Mountains, United States | 18 | Sea of Cortez (Gulf of California). River rarely reaches the sea. | Delta and sea ecosystems are deteriorating; salt levels are rising; fisheries are declining; many species are threatened; indigenous culture is at risk. |

Source: Global Water Policy Project and Worldwatch Institute, based on a variety of sources.

so essential to Egypt's economy, is slowly falling into the sea. Most deltas sink from the weight of their own sediment, but under natural conditions this subsidence is usually countered by deposition of silt brought in by the river. The Nile transports an average of 110 million tons of silt each year, much of it fertile soil washed down from the Ethiopian highlands. For thousands of years, 90 percent of this silt reached the coast to replenish the delta, while the remaining 10 percent was deposited on the Nile floodplain. The delta stopped growing about a century ago, after the first barrages (small dams) were built by the British. But since completion of the High Dam at Aswan, and the trapping of virtually all of the silt in Lake Nassar, the delta has actually been in retreat. Borg-el-Borellos, a former delta village, is now 2 kilometers out to sea.[45]

Global warming, and the anticipated rise in sea level that higher temperatures will bring, increases the threat of inundation. Much of the northern delta lies only 3-4 meters above sea level. Researchers at the Woods Hole Oceanographic Institution in Massachusetts calculate that Egypt could lose up to 19 percent of its habitable land within

about 60 years, displacing up to 16 percent of its population—which by then would likely total well over 120 million—and wiping out some 15 percent of its economic activity.[46]

No region better illustrates the consequences of under-
valuing ecosystem services than the Aral Sea basin in central
Asia. Some four decades ago, Soviet central planners calculated that using central Asian rivers for irrigation of cotton
would produce more economic value than letting the majority of their flow empty into the Aral Sea, which was then the
planet's fourth largest lake. Irrigated area in the region
expanded greatly during the ensuing decades, and now
totals 7.9 million hectares. This places the Aral Sea basin
among the world's largest irrigation systems, with an area
more than double that of Egypt's and half that of the vast
Indus system in Pakistan.[47]

Prior to 1960, the Amu Dar'ya and Syr Dar'ya poured
55 billion cubic meters of water a year into the Aral. As river
diversions for irrigation increased, however, this flow diminished. Between 1981 and 1990, the rivers' combined flow
into the sea dropped to an average of 7 billion cubic meters,
13 percent of the pre-1960 inflow. (See Figure 3.) Like a big
swimming pool in the desert, the Aral needs substantial
replenishment to counter the large volume of water it loses
to evaporation. Without that replenishment, the sea
shrinks. To date, the Aral has lost half its area and three-
fourths of its volume. Unusually heavy rains in the Aral
watershed between 1990 and 1994 increased annual river
inflow to an average of 23 billion cubic meters; but even this
was not enough to stop the sea's shrinkage.[48]

The still-unfolding chain of ecological destruction
ranks the Aral Sea's demise as one of the planet's greatest
environmental tragedies. Twenty of the 24 fish species there
have disappeared, and the fish catch, which totalled 44,000
tons a year in the 1950s and supported 60,000 jobs, has
dropped to zero. Abandoned fishing villages dot the sea's
former coastline. Each year, winds pick up some 100 million
tons of a toxic dust-salt mixture from the dry sea bed and

FIGURE 3

River Flow into the Aral Sea, 1940–90

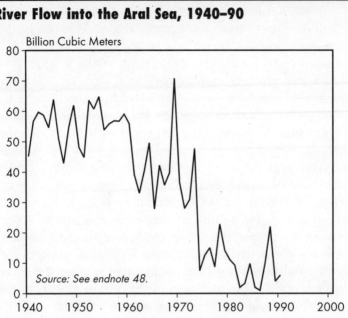

Billion Cubic Meters

Source: See endnote 48.

dump them on the surrounding farmland, harming or killing crops. The low river flows have concentrated salts and toxic chemicals, making water supplies hazardous to drink and contributing to high rates of many diseases. The population of Muynak, a former fishing town, is down from 40,000 several decades ago to just 12,000 today. The 28,000 people who have fled are "ecological refugees" in the truest sense.[49]

Both the Amu Dar'ya and Syr Dar'ya deltas have been severely degraded by the diminished river flow. According to Philip Micklin of Western Michigan University, a leading U.S. authority on the Aral Sea basin, the tugay forests of willow, tamarisk, and other water-loving trees and shrubs that are vital habitat for the region's animal life have been decimated. Wetlands have shrunk by 85 percent, which, combined with high levels of agricultural chemical pollution, has greatly reduced waterfowl populations. In the Syr Dar'ya Delta, the number of nesting bird species has fallen from an estimated 173 to 38.[50]

What has happened in the Aral Sea basin shows vividly how damage to economy, community, and human health can follow close on the heels of ecological destruction. It is a lesson worth heeding in other parts of the world where large water projects are planned or underway. In Latin America, a scheme known as the Hidrovia (Portuguese for waterway) would channelize the Paraguay-Paraná rivers in order to link the heart of the continent with Atlantic Ocean ports. If carried out, the scheme could destroy a portion of the ecological values of the Pantanal—the world's largest intact area of wetlands, a valuable source of natural flood protection for the region, and home to 600 species of fish, 650 species of birds, and 80 species of mammals. Major work on the project is not supposed to begin until environmental assessments are completed. Individual countries in the five-nation consortium behind the scheme, however, have announced intentions to begin certain aspects of the work, which could threaten wetlands in Paraguay and risk making some version of the Hidrovia a fait accompli before the environmental studies are even completed by the end of 1996.[51]

In a few cases where large schemes have led to unacceptable consequences, major restoration efforts are being undertaken. In the United States, federal and state agencies have given the green light to a large-scale effort to revitalize south Florida's Everglades ecosystem, the famed "river of grass" and treasure trove of wildlife that shrank by half as the natural water system was manipulated for agricultural and urban expansion. At a cost of up to $2 billion over the next 15-20 years, the effort aims to restore the water rhythms of the original natural system. Restoration work is also underway in the Aral Sea basin under a complex, multi-donor program coordinated by the World Bank. If political and financial commitments allow them to proceed, the Aral basin and Everglades efforts will be two of the largest ecosystem restoration schemes ever undertaken, and they will be closely watched as potential models. They are expensive, however, and offer no guarantee of success. At bottom, their costliness and complexity testify to the need to ensure that, from the outset, economic development and water engineering do not compromise the health and functioning of aquatic ecosystems.[52]

Competition and Conflict

A third major threat to human security arises from the heightening competition for water both within and between countries as supplies increasingly fall short of needs. A new politics of scarcity is emerging as farms and cities, states and provinces, and neighboring countries compete for a limited or shrinking supply.

Three principal forces conspire to create scarcity and its potential to incite conflict or dispute: the depletion or degradation of the resource, which shrinks the "resource pie"; population growth, which forces the pie to be divided into smaller slices; and unequal distribution or access, which means that some get larger slices than others. Although all three often play a part, it appears that unequal distribution often has the most important role.[53]

In the Jordan River basin, for example, Israel strictly limits Arab access to groundwater in the occupied West Bank. On a per capita basis, Israeli settlers there use about four times more water than neighboring Arabs. Since 1967, when Israeli occupation of the West Bank began, Palestinians have had to obtain permission from Israeli military authorities to drill new wells. As of 1991, permission had been granted only 19 times, and only three of the approved wells were for agricultural use. Meanwhile more than 30 agricultural wells have been drilled for Israeli settlers, and domestic supplies are often sufficient to fill swimming pools. Adding to the inequity, the Palestinians pay $1.20 per cubic meter for piped water, which is triple what Israeli settlers pay for domestic water and 7.5 times what they pay for irrigation water. While the links are not clear-cut, it seems likely that inequitable water rights and access contributed to the anger underlying the *intifada* (Palestinian uprising).[54]

In the interim agreement signed in late September 1995 for the continued transfer of authority to the Palestinians, Israel recognized, for the first time, that the Palestinians have rights to West Bank groundwater. That agreement also called for a joint Israeli-Palestinian committee to manage water in the West Bank. And during the interim phase of the peace process, 70-80 million cubic meters of water is to be made available to Palestinians annually, at cost, for any purpose. While these are positive steps toward a peaceful resolution of the water dispute (although follow-through is now uncertain as a result of the change in Israeli leadership), much remains to be done. The interim allotment to the Palestinians, for example, is likely to be sufficient to cover only household needs. Also unresolved is the larger issue of the permanent allocation of rights to West Bank water (which now accounts for 25 percent of Israel's total nationwide supply) and to other waters of the Jordan basin.[55]

In some cases, dams and other development projects intended to improve conditions for agriculture or the economy can end up fueling tensions if newly created access to the scarce resource worsens existing inequalities, further marginalizes the poor, or creates opportunities for the rich to "capture" the resource. In the Senegal River basin, for example, farming, herding, and fishing traditionally depended on the river's yearly cycle of flooding. In the 1970s, concern about chronic food shortages and drought led governments in the region to seek financing for the Manantali Dam in order to expand irrigated agriculture, hydropower production, and river transport. Wishing to capture the benefits that new irrigation water would bring to the land adjacent to the river, the Mauritanian elite—mainly consisting of white Moors—prevented black Africans from continuing their flood-based activities along the river.

Ethnic violence broke out in both Senegal and Mauritania, with the Moors forcing 70,000 black Mauritanians to move into Senegal.[56]

In the water-short Jodhpur district of Rajasthan, India, village wells tapped by the rural poor are drying up as deeper wells to supply the cities have caused water tables to drop, and as groundwater has increasingly been used on chili peppers and other water-intensive crops grown for commercial sale. As a result of their loss of access to common water sources, poor villagers cannot get the food and fodder they need. To sustain themselves, the men take on health-threatening work as low-wage laborers in the quarry mines, and the women spend an average of four hours a day collecting water. As researcher Michael Goldman points out, "Peasant families have lost their access to groundwater and have had to watch their herds die, land deteriorate, and their families and communities split up....The privatization of the rural commons is intensifying exploitative social relations and degrading ecological relations."[57]

On a per capita basis, Israeli settlers in the West Bank use about four times more water than neighboring Arabs.

Just as tensions over water are mounting between ethnic or societal groups, the potential for hostility and conflict between countries is rising as well. Unique among strategic resources, water flows easily across political boundaries. Many countries depend on river water from upstream neighbors for a substantial portion of their surface supplies. (See Table 6.) Particularly in the face of population growth and rising water demands, these countries can become highly vulnerable to decisions by upstream countries to siphon off more water for themselves. According to Thomas Homer-Dixon of the University of Toronto, codirector of the Project on Environmental Change and Acute Conflict, the evidence suggests that "the renewable resource most likely to stimulate interstate resource war is river water."[58]

River basins in which hostilities are most likely to erupt are those in which the river is shared by at least two countries, water is insufficient to meet all projected demands, and there is no recognized treaty governing the allocation of water among all basin countries. Examples of such potential hot spots include the Ganges, the Nile, the Jordan, the Tigris-Euphrates, and the Amu Dar'ya and Syr Dar'ya.

Outright conflict has the greatest potential to emerge when the downstream (most vulnerable) nation is militarily stronger than the upstream (water-controlling) nation and feels that its interests are threatened. Prior to 1967, for example, Israel was in a disadvantageous position with regard to water, but was more powerful than its immediate neighbors. Syrian attempts to divert the Banias, one of three sources of the upper Jordan River (which, in turn, feeds the Sea of Galilee, Israel's principal source of surface water), contributed to rising tensions and a series of armed confrontations with Israel immediately preceding the Six Day War in 1967. Israel's victory in that conflict gave the nation control over two areas of strategic water importance—the West Bank mountain aquifer and the Golan Heights, which feeds the Banias into the upper Jordan and also provides access to the site of an intended Jordanian dam on the Yarmouk River.[59]

Egypt, which gets hardly any rainfall, is perhaps more vulnerable than any other country to a reduction in water supplies. The nation depends on the Nile River flowing into its territory for 97 percent of its surface water. With a population of 60 million climbing by 1 million every nine months, some 3.2 million hectares of cropland totally dependent on irrigation, and a current water demand that is very near the limits of the supply, any cutoff of Nile flow would be highly disruptive, if not disastrous. University of Pennsylvania professor Thomas Naff noted in 1994 that "it is an axiomatic policy of every Egyptian regime that it will go to war, if necessary, to prevent either of its closest upper riparian neighbors, Sudan and Ethiopia, from reducing in any way the flow of the Nile."[60]

Until recently, Egypt was at minimal risk of suffering

TABLE 6

Dependence on Imported Surface Water, Selected Countries

Country	Share of Total Flow Originating Outside of Border (percent)
Turkmenistan	98
Egypt	97
Hungary	95
Mauritania	95
Botswana	94
Bulgaria	91
Uzbekistan	91
Netherlands	89
Gambia	86
Cambodia	82
Syria	79
Sudan	77
Niger	68
Iraq	66
Bangladesh	42
Thailand	39
Jordan	36
Senegal	34
Israel[1]	21

[1]Includes only flows originating outside current borders; a significant additional share of Israel's fresh water originates from occupied, disputed territories.
Source: See endnote 58.

such reductions, except, of course, from drought. But Ethiopia, where 86 percent of the Nile's total flow originates, now has the political stability and capacity to mobilize resources to store and use water for agricultural and economic advancement. An estimated 3.7 million hectares of Ethiopia's land is potentially irrigable. Using Nile water to irrigate even half this area could reduce downstream flows by some 9 billion cubic meters per year—equal to 16 percent of Egypt's current annual Nile supply. Moreover, Ethiopia plans to expand hydropower production, with some 80 percent of future hydro schemes located on Nile tributaries. Similarly, studies of Uganda, in the upper White Nile basin, suggest that on the order of 2 billion cubic meters per year of additional water consumption might occur there if its irrigation potential were fully developed. Thus Egypt seems increasingly vulnerable to a loss of Nile water.[61]

A somewhat similar situation exists in the Aral Sea basin. Afghanistan, Iran, and five countries newly independent after the breakup of the Soviet Union—Kazakhstan, Kyrgyzstan, Tajikistan, Turkmenistan, and Uzbekistan—form the basin and share the waters of the Amu Dar'ya and Syr Dar'ya. In addition to having to deal with the destruction of the Aral Sea ecosystem and its consequences, these countries face fundamental water challenges: there is not enough water in the basin to meet all demands. Disputes have already occurred between Kyrgyz and Uzbeks over water and land in the Fergana area, between Kyrgyz and Tajiks over the allocation of irrigation water, and between Turkmens and Uzbeks over the distribution of irrigation and drainage water in the Amu Dar'ya Delta. At the moment, larger-scale conflict appears unlikely since these five countries—which account for the vast majority of the basin's water use—continue to use the water-allocation formula set by Moscow. The status quo, however, is neither equitable nor environmentally sustainable, and, as described in the next section, efforts are underway to arrive at more acceptable water-sharing arrangements.[62]

When downstream countries are relatively less powerful than water-controlling upstream countries, conflict may be less likely, but social and economic insecurity—which in turn can lead to political instability—can be great. For example, as the weaker riparian, Bangladesh will almost certainly not choose to go to war with India. But as the nation last in line to receive water from the Ganges—which rises in the Himalaya of Nepal and then flows through India and Bangladesh before emptying into the Bay of Bengal—

Bangladesh is losing out, and the failure to meet its needs is having a destabilizing effect on relations with its more powerful neighbor.

In the early 1970s, India completed the Farakka Barrage to divert Ganges water to the port city of Calcutta, which reduced the flow into Bangladesh. The two nations agreed in 1977 to a short-term solution for sharing the dry-season flow, and also guaranteed Bangladesh a minimum amount of water during extremely low-flow periods. That agreement expired in 1982 and was replaced with an informal accord that did not include the guarantee clause for Bangladesh. A follow-up agreement expired in 1988. Since then, the two countries have been deadlocked, leaving Bangladesh with no assurance of minimum flows for its dry-season irrigation needs.[63]

Eighty-six percent of the Nile's total flow originates in Ethiopia.

Tensions between the two countries have worsened in recent years. In 1993, the dry-season flow into Bangladesh was the lowest ever recorded. As river beds dried up and crops withered, the western region suffered greatly. The Ganges Kobadak project, one of this poor nation's largest agricultural schemes, reportedly suffered an estimated $25 million in losses. Irrigation pumps on the Gorai River, the Ganges' main tributary in Bangladesh, were idle again in 1994. And in October 1995, then Prime Minister Begum Khaleda Zia stated before the United Nations that more than 40 million Bangladeshis were facing poverty and suffering because of India's diversions of Ganges River water. She called India's actions "a gross violation of human rights and justice," and said the Farakka Barrage had become for Bangladeshis "an issue of life and death."[64]

Syria and Iraq are in a similar situation with regard to Turkey, the eastern mountains of which give rise to both the Tigris and Euphrates rivers. Turkey is undertaking a huge hydropower and irrigation scheme known as the GAP (after the Turkish acronym), which could reduce the Euphrates flow into Syria by 35 percent in normal years and substantially more in dry ones, besides polluting the river with irrigation drainage. Iraq, third in line for Euphrates water, would see a reduction as well. Syria and Iraq have agreed to share whatever mainstream Euphrates water crosses the border into Syria from Turkey, with Syria getting 42 percent and Iraq getting 58 percent.[65]

Turkey and Syria signed a protocol in 1987 that guarantees the latter nation a minimum flow of 500 cubic meters per second, about half of the Euphrates' volume at the border, but Syria wants more—a request Turkey so far has denied. In 1992, then Turkish Prime Minister Suleyman Demirel remarked about Syrian requests for more Euphrates water: "We do not say we should share their oil resources. They cannot say they should share our water resources." Although the government may have a more compromising position than this hard-line language would suggest, the parties have not yet achieved a water-sharing agreement.[66]

Sharing the Waters

Historically, rivers have often been used to delineate political boundaries, and have thus divided nations. Ecologically, however, rivers join nations. Any river that forms a border between two countries courses through the middle of a watershed that spans those two countries. And any river that flows through two or more nations—as 214 do—is supported by ecosystems that cut across political boundaries. Cooperation is thus essential not only to avert conflict but to protect the natural systems that underpin regional economies. It is especially critical now that much of the world has entered a zero-sum game in which increasing the water available to one user means taking some away from another. Whether in the marketplace or in international politics, allowing competition alone to sort out winners and losers is a no-win proposition for all; in today's interdependent world, the spoils of victory would soon be offset by the costs of regional instability and ecological decline. As the basis of life, water requires an ethic of sharing—both with nature and each other.[67]

Although no enforceable law governs the allocation and use of international waters, a code of conduct and legal framework for shared watercourses has steadily been evolving. The private International Law Association (ILA) provided guidelines for sharing common waterways in its Helsinki Rules of 1966, which state: "Each basin State is entitled, within its territory, to a reasonable and equitable share in the beneficial uses of the waters of an international drainage basin." The rules then list about a dozen factors that should be taken into account in determining what is reasonable and equitable, but offer no guidance on prioritizing or weighting them.[68]

A second body, the United Nations International Law Commission (ILC), presented its set of "Rules on the Non-Navigational Uses of International Watercourses" to the U.N. General Assembly in 1994. Like the ILA's guidelines, the ILC's rules are intended to provide general principles that can then be applied to specific river basins. The most recent set of rules gives clear priority to the "equitable and reasonable" use principle over that of the "obligation not to cause significant harm," the latter having been disliked by upstream countries that feared it would hamstring their water development activities. The ILC rules also list factors to be taken into account in determining what constitutes equitable and reasonable use, but, like the Helsinki rules, assign no weighting or priority to them.[69]

Establishing equity and reasonableness as the overriding principle for water allocation and use in shared river basins is a clear contribution of international law, but the vagueness of the principle makes it only minimally helpful in practice. Determining what constitutes "reasonable and equitable" is the crux of any water-sharing agreement, and it is open to widely differing interpretations. For instance,

Egypt's view of such an allocation of the Nile would undoubtedly give great weight to population size and historical water use. Ethiopia, on the other hand, would place relatively greater emphasis on each nation's contribution to total runoff in the basin and on future irrigation potential.

In the absence of a formal body of practical and enforceable law, water sharing and the prevention of conflict depend on treaties between the parties involved. Governments have concluded more than 2,000 legal instruments relating to international watercourses, with some dating back 900 years. Most of the treaties that set forth allocations of water quantity or quality reflect the basic principle of equitable use, even if they do not use that language. Unfortunately, however, in none of today's potential hot spots of water dispute does a treaty exist that includes all parties within the river basin.[70]

The 1994 treaty signed by Israel and Jordan, for example, resolves some of the water issues between these two countries, and hopeful early signs of a peaceful resolution of water disputes are evident in the 1995 Israeli-Palestinian agreement as well. But until water rights or allocations are clarified and agreed to by all parties in the Jordan River basin—which include Israel, Jordan, Lebanon, the Palestinians, and Syria—tensions will likely persist. In an April 1996 speech in which U. S. Secretary of State Warren Christopher described a "vast new danger posed to our national interests by damage to the environment and resulting global and regional instability," he defined the regional element of a new foreign policy strategy as confronting "pollution and the scarcity of resources in key areas where they dramatically increase tensions within and among nations." He went on to single out "the parched valleys of the Middle East, where the struggle for water has a direct impact on security and stability." Having the U.S. Secretary of State publicly make these linkages is a highly promising development, but it remains to be seen whether a U.S. initiative will follow that gives impetus to a lasting, basinwide solution to the region's water disputes.[71]

In the Nile basin, a 1959 treaty between Egypt and Sudan allocates an amount of Nile water between them that adds up to nearly 90 percent of the river's average annual flow—even though 86 percent of that flow originates in Ethiopia. Ethiopia is not party to the treaty and, not surprisingly, feels no obligation to respect it. Fortunately, now that Ethiopia is in a position to begin tapping upper Nile waters for its own use, the affected nations are beginning to cooperate. An intergovernmental organization called Tecconile is focusing on technical aspects of water resource development in the basin; from its headquarters in Uganda, it serves all member and observer countries on an equal basis. At a February 1995 meeting in Tanzania, the water affairs ministers of most of the Nile basin countries—including Egypt and Ethiopia—agreed to form a panel of experts that would be charged with developing a basinwide framework for water sharing aimed at "equitable allocation of the Nile waters." Especially given Egypt's historic position, this is a striking development, one that may not only avert conflict over water in the Nile basin, but also set the stage for more sustainable water management and use.[72]

In the Aral Sea basin, the presidents of Kazakhstan, Kyrgyzstan, Tajikistan, Turkmenistan, and Uzbekistan met in January 1994 and approved an action plan for addressing the basin's dire situation and for broader social and economic development over the next three to five years. The centerpiece of the plan is a regional water management strategy, completed in draft form in May 1996, which has been agreed to by all five countries. In several key respects, this strategy document is a milestone for basin-wide water management. It recognizes the Aral Sea and the Amu Dar'ya and Syr Dar'ya delta ecosystems as "water users" in their own right, deserving of water allocations. It confirms that principles of international law should apply to decisions about interstate water allocation.

No enforceable law governs the allocation and use of international waters.

And it acknowledges the unsustainability of current agricultural practices in the basin.[73]

Carrying out the broad water, agricultural, economic, and social reforms that will be needed to achieve sustainable water use in the Aral Sea basin, however, will be difficult, to say the least: the economies of each of the five countries contracted by 10 to 55 percent between 1990 and 1994. The strategy reflects an understandable reluctance to effect radical, overnight change in water management policies and procedures. And some of the nations' economic goals—such as continued irrigation expansion—run counter to that of securing more water for the Aral Sea and delta ecosystems. Nonetheless, there is now clear potential in the Aral Sea basin, not only for strengthened cooperation, but for a new basin-wide framework that promotes more sustainable patterns of water use.[74]

By contrast, little progress is apparent in the Ganges basin, where a water-sharing agreement between Nepal, India, and Bangladesh becomes more urgent each year. With Nepal holding most of the additional water storage and hydropower potential, India controlling the vast majority of Ganges runoff, and Bangladesh repeatedly capturing international attention with its debilitating floods and droughts, this basin is ripe for creative initiatives. Initial steps might include an independent or trilateral panel to develop "reasonable and equitable" arrangements that secure Bangladesh's minimum water needs for food production as well as the minimum water needed to safeguard the Ganges Delta. With increasing numbers of Bangladeshi refugees crossing the border into eastern India to escape poverty, an outmigration at least partially propelled by lack of water, India may soon see reason to bargain where it saw little before.[75]

A number of lessons emerge from past and ongoing efforts to arrive at international water-sharing agreements. First, some third-party involvement is often key to resolving water disputes, and this involvement may need to be backed by financial support. For example, the World Bank played a key intermediary role in resolving the 12-year dispute between India and Pakistan over the Indus River that erupted in 1947 with the partitioning of the subcontinent. In addition, the Bank's mobilization of financing for carrying out technical aspects of the agreement was integral to the success of the Indus Waters Treaty, which was signed by the two countries in 1960. The World Bank is also playing a key coordinating role in the Aral Sea basin, and is helping to mobilize funding from donor countries for a wide variety of projects there.[76]

Second, water agreements may be easier to achieve, and economically more efficient, if they include resources or assets other than water. A natural trade in many river basins, for example, is that between water supply and energy. An agreement signed by three of the Aral Sea basin countries in April 1996 calls for Kyrgyzstan to guarantee hydroelectric power and sufficient Syr Dar'ya flow for irrigation downstream in Uzbekistan and Kazakhstan in return for some of Uzbekistan's natural gas and Kazakhstan's coal. One idea for resolving the dispute over the Ganges is for India to agree to increase the dry-season flow into Bangladesh in exchange for transit rights across the latter's territory to its own northeastern states. The potential for creative linkages such as these is virtually limitless. Equity would dictate, however, that a certain minimum amount of water be provided to all parties in a given river basin, and that no party should have to trade other assets to receive this minimum amount.[77]

Third, progress toward water-sharing agreements is sometimes made when the negotiations shift from discussions of water rights to water needs. How much water each party has a "right" to is subjective, emotionally charged, and varies with the criteria used, but how much each "needs" or can beneficially use can more easily be quantified objectively. For example, the Johnston Accord for the Jordan River basin, which was orchestrated in the 1950s, but never ratified for political reasons, took a needs-based approach. Specifically, it involved estimating how much water was

needed for all the potentially irrigable land within the basin
that could receive water by gravity flow. National water
allocations were then determined by the location of this
land within the basin. Although irrigable area was the basis
for determining the allocation, each country could use its
share of water any way it pleased. The Johnston water-sharing formula was acceptable to all parties at the time and still
has validity today.[78]

Fourth, an ongoing project at Harvard University suggests that monetizing the disputed water may lessen the
emotional charge of the dispute, and thus pave the way for
an agreement. The Harvard team has estimated, for example, that the monetary value of the water in contention in
the Jordan basin presently totals no more than $110 million
per year—not an amount, the researchers suggest, that
should cause war. They claim that their model can encourage efficient water use and, by providing the parties with a
useful policy tool, "can also lead to a concentration on joint
benefits, removing the belief that a zero-sum game is being
played."[79]

The inevitability of droughts and the prospect of climate change must also figure into water-sharing agreements. It may no longer make sense for treaties to specify
the absolute quantity of water each nation, state, or
province receives, since in many years there may not be
enough water to meet all treaty requirements. A more sen-
sible approach is for agreements to specify each party's
respective share of river runoff, with the absolute amount
each gets tied to how much is available in a particular year.
To protect a river's ecological functions, treaties would need
to specify an absolute quantity and quality of water that is
reserved for the environment, and this minimum flow
would need to be provided in dry years or wet.

The Murray-Darling river basin in Australia is now
managed under such an approach. During the last decade,
as water use in the basin approached the sustainable yield of
available water resources, pressure increased to better define
ownership of water supplies among the four basin states—

Queensland, New South Wales, Victoria, and South Australia—and to share them more equitably, particularly during droughts. The Murray-Darling River Basin Commission, which works across state borders with government departments and community representatives to manage the basin's water resources, assesses the quantity of water available for use each year. Before making any allocations to the basin states, the Commission determines minimum river flows for ecosystem health and establishes reserve storage to safeguard against future droughts. The remaining water is then shared among the states according to proportions set forth in the 1987 Murray-Darling Basin Agreement. The system includes continuous accounting of water use in the basin states and, for added flexibility, allows for water trading among the states.[80]

It may no longer make sense for treaties to specify the absolute quantity of water each nation, state, or province receives.

Finally, creating institutions and procedures that allow for joint, integrated management of water that crosses political boundaries is critical. Sharing water equitably is only one part of the challenge in water-short river basins: using and managing the water optimally is another. When countries in the same river basin are cooperating and managing the basin's water in an integrated, holistic manner, a host of strategies become feasible that are simply impossible when they manage water separately and in piecemeal fashion. Such cooperation is not easy to achieve, however, because countries are typically reluctant to give up sovereignty and independence in their water management decisions. The signing of the Indus Waters Treaty by India and Pakistan, for example, is a widely lauded success story because it resolved the long-standing water dispute between the two countries and came up with a formula for sharing the Indus and its tributaries—but it did not result in joint, integrated management of the basin's waters.

The potential benefits and win-win possibilities offered by joint river basin management, however, may nudge reluctant countries to try it as water becomes increasingly scarce. In the Nile basin, for example, 12 percent of the river's average flow is lost through evaporation from Lake Nassar each year. Storing more Nile water in the Ethiopian highlands, where evaporation levels are one-third those at Aswan, could reduce these losses and thus increase the amount of water available for all the Blue Nile countries. This, in turn, might allow Ethiopia to expand irrigation without reducing Egypt's current water supplies, thereby eliminating a potential source of conflict between the two countries. Recent signs of cooperation in the basin give some hope that such joint management strategies might eventually be pursued, but substantial institutional development and trust-building would need to take place first. Likewise, joint management is critical in the Aral Sea basin. Here, the formation of the Interstate Council for the Aral Sea and a number of other new institutions to oversee water and related environmental matters among the basin countries is a promising step forward.[81]

Priority Actions

P reventing water scarcity from undermining food security, ecological life-support systems, and social and political stability will not be easy. In much of the world, expanding the water supply to one user now means taking it away from another. New dams and river diversions will rarely offer sustainable solutions, since in most cases they would involve taking more water from freshwater systems that are already overtaxed. The key challenges now are to establish priorities and policies for allocating water among competing uses and users, to encourage more efficient and productive use of water, and to reshape institutions to better suit the new era of water constraints. These are not challenges that water managers can meet alone. They now belong in the portfolios of diplomats, on the agendas of cabinet meetings, and high on the priority lists of development banks and international support agencies.

A top priority is to ensure that both people and ecosystems get at least the minimum amount of good-quality water they need to remain healthy and to function productively. Especially with competition for scarce water increasing and strong pressures to treat water more as a commodity, governments have an important responsibility to ensure that water's most fundamental function—supporting life—is fulfilled.

Studies suggest that on the order of 25 liters of water per person per day are needed for a survival level of drinking water and adequate sanitation; adding in the water required for good hygiene and food preparation would bring the minimum amount to about 50 liters per person per day. Today, more than a billion people lack even this. Combining the minimal requirement of these people with that of the 2.6 billion people projected to be added to world population by 2025 yields a total additional water requirement for human health of some 65 billion cubic meters annually by 2025—equal to just 1.5 percent of current water extractions worldwide. Satisfying these basic human needs is thus not constrained by water availability per se, but rather by inadequate investment by governments, external support agencies, water providers, and community groups in the technologies, infrastructure, and institutions needed if the poor are to have access to safe water sources. Tariff structures that provide this minimum amount at "lifeline rates" (which permit payments the poor can afford for amounts of water essential to their survival), perhaps subsidized by high prices for luxury levels of consumption, can help satisfy basic needs while allowing recovery of system costs.[82]

Protecting the health and functioning of freshwater ecosystems has gotten much less attention than the challenge of providing safe drinking water and sanitation services. Determining how much water—and of what quality—is needed to keep ecosystems in good working order is more complicated than determining the minimum amount required for human health. Exactly how much water needs to be left in a river will vary with the time of year, the habitat requirements of riverine life, the system's sediment and salt balances, the value local residents place on fisheries and recreation, and other factors specific to each river basin. But setting even preliminary minimum flows for both average and low-flow periods would provide some needed assurance of ecosystem protection. These levels can then be refined as more knowledge is gained about river system functioning.

In the many parts of the world where rivers already are overtapped, meeting the water needs of ecosystems will require shifting some water away from farms and cities. A few such efforts have been initiated in the United States. In late 1992, for example, the U.S. Congress passed legislation that dedicated 800,000 acre-feet (987 million cubic meters) of water annually from the Central Valley Project in California, one of the largest federal irrigation projects, to maintaining fish and wildlife habitat and other ecosystem functions. Among other aims, it set a goal of restoring the natural production of salmon and other anadromous fish (those that migrate from salt water to fresh water to spawn) to twice their average levels over the past 25 years. Efforts are also underway to limit the amount of fresh water that can be diverted from the San Francisco Bay delta-estuary, a highly productive aquatic ecosystem that is home to more than 120 species of fish. And a 1994 California Supreme Court decision mandated that Los Angeles reduce its withdrawals from tributaries feeding Mono Lake, which had dropped 13.7 meters and shrunk in volume by half over several decades because of the city's diversions. In the two years since the decision, which was based on a broader interpretation of a legal doctrine called the "public trust," the lake has already risen nearly 1.8 meters.[83]

Allocating water to the environment may be more difficult in developing countries, where water demands are rising rapidly because of high rates of population and economic growth. But in these countries as well, ensuring minimum water flows for natural ecosystems is critical to protecting fisheries, delta economies, and the health of local people. Part of the World Bank-coordinated program in the Aral Sea basin involves constructing wetlands and artificial lakes in the Amu Dar'ya Delta in order to restore aquatic vegetation, fisheries, and wildlife. But the Aral Sea ecosystem would need a substantial allocation of water just to stop the spiral of decline, much less reverse it. Stabilizing the sea even at its present level would require an annual inflow of some 35 billion cubic meters—five times greater than the average annual inflow registered during the 1980s. Shifting this much water back to the ecosystem would take major irrigation efficiency improvements, a reduction in the area planted to cotton and rice, and the removal of marginal lands from irrigation.[84]

Nearly the entire spectrum of conservation and efficiency options cost less than the development of new water sources.

Ecosystems would be much more likely to receive protection and a dedicated water allocation if policymakers attached economic value to the services they provide. Wetlands, for example, offer flood protection, water purification, and habitat benefits that are "public goods" for which no one specifically pays a price. As a result, these ecosystem services are not adequately taken into account by project planners and decisionmakers, and thus are lost or destroyed at a more rapid rate than is optimal. For example, the value of coastal Malaysian mangrove swamps for flood control and storm protection alone has been placed at $300,000 per kilometer. Yet, in much of Southeast Asia and coastal Latin America, mangrove forests are rapidly being destroyed for shrimp farming and other activities of lesser and shorter-lasting value. Similarly, a number of studies of the western United States suggest that, at least during periods of low river flow, the "marginal value" of water for habitat protection and for fishing, boating, and other instream recreational uses equals or exceeds that of water used in a substantial portion of irrigated agriculture. By attaching monetary values to ecosystem services—even if they are only roughly accurate—the more serious mistake of implicitly valuing these services at zero is avoided. Only in this way are ecosystems likely to receive protection and water allocations commensurate with their actual worth.[85]

Once sufficient water has been assured for human and ecosystem health, the challenge is to establish policies that encourage more efficient and productive use of the water remaining. Stretching existing water supplies can help satisfy additional urban, industrial, or irrigation needs within countries, as well as relieve tensions between countries. Moreover, when compared on an equal footing with water supply projects, measures to reduce the demand for water through investments in conservation, recycling, and increased efficiency are typically the most economic alternatives for balancing water budgets. At 5-50¢ per cubic meter, nearly the entire spectrum of conservation and efficiency options—including leak repair, the adoption of more efficient technologies, and water recycling—cost less than the development of new water sources. (See Table 7.) Even the most expensive conservation options cost half as much as the least expensive seawater desalination projects, and the lower-end conservation options cost 5-10 percent as much as seawater desalination.[86]

Unfortunately, large subsidies to water users continue unchecked, discouraging efficiency investments and conveying the false message that water is abundant and can be wasted—even as rivers are drying up, aquifers are being depleted, and fisheries are collapsing. Farmers in water-short Tunisia pay 5¢ per cubic meter for irrigation water—one-seventh the cost of supplying it to them. Jordanian farmers pay less than 3¢ per cubic meter, a small fraction of the water's full cost. Federal construction cost subsidies to irrigators in the western United States are estimated to total

TABLE 7

Estimated Costs of Water Management Options, c. 1995

Management Option	Estimated Cost Range (cents per cubic meter)
Reducing demand through conservation/efficiency	5-50
Treatment and reuse of wastewater for irrigation	30-60
Desalination of brackish water	45-70
Development of marginal water sources	55-85
Desalination of seawater	100-150

Source: World Bank, From Scarcity to Security: Averting a Water Crisis in the Middle East and North Africa (Washington, D.C., 1995).

at least $20 billion, representing 86 percent of total construction costs. And in India, the Madras Institute of Development Studies estimates that less than 10 percent of the total recurring costs for major- and medium-sized irrigation projects built by the government as of the mid-1980s had been recovered.[87]

Although there may be sound social reasons to subsidize irrigation to some degree, especially for poor farmers, the degree of subsidization that exists today is a major barrier to achieving more sustainable water use. Charging prices that at least cover operation and maintenance costs and send a proper signal about the need for efficiency improvements is essential. There is a broad spectrum of options between full-cost pricing, which could put many farmers out of business, and a marginal cost of nearly zero to the farmer, which is a clear invitation to waste water.

In the Broadview water district in California, for example, farmers irrigate 4,000 hectares of melons, tomatoes, cotton, wheat, and alfalfa. In the late 1980s, when the district was faced with the need to reduce drainage into the San Joaquin River, it established a tiered water pricing structure. The district determined the average volume of water used over the 1986-88 period, and applied a base rate of $16 per acre-foot (1.3¢ per cubic meter) to 90 percent of this

amount. Any water used above that level was charged at a
rate 2.5 times higher. In 1991, only 7 out of the 47 fields in
the district had any water charged at the higher level, the
average amount of water applied on the district's farms had
dropped by 19 percent, and salt releases had fallen by 4,000
tons. Even though farmers were still paying prices far below
the water's real cost, the pricing structure nonetheless created an incentive for them to conserve.[88]

Because agriculture accounts for two-thirds of total
water use worldwide, even small-percentage reductions can
free up substantial quantities of water for cities, the environment, or additional food production. Savings of up to 25
percent or more from efficiency improvements are well doc-
umented for certain locations. Farmers in northwest Texas,
for example, who have had to cope with falling water tables
from depletion of the Ogallala aquifer, have reduced their
water use by 20-25 percent by adopting more efficient sprinkler technologies, surge valves to even out distribution in
gravity systems, and other water-saving practices.[89]

Likewise, results from a variety of countries show that
farmers who have switched from furrow or sprinkler irrigation to drip systems, which deliver water directly to the
roots of crops, have cut their water use by 30-60 percent.
Yields often increase at the same time because plants are
effectively "spoon-fed" the optimal amount of water (and
often fertilizer) when they need it. Drip systems, which cost
in the range of $1,200-$2,500 per hectare, tend to be too
expensive for most poor farmers and for use on low-value
row crops, but research is underway to make them more
affordable. Colorado-based International Development
Enterprises (IDE) has developed a drip system that costs just
$50 per half acre ($123 per half hectare), 10-20 percent of
the cost of traditional drip systems. The keys to keeping
costs down are simple materials and portability: instead of
each row of crops getting its own drip line, a single line is
rotated among ten rows. In field trials in a hilly region of
Nepal, the system doubled the amount of land that could be
irrigated with the same volume of water.[90]

Just how much water can be saved through efficiency improvements will vary from place to place. Worldwide, irrigation efficiency is estimated to average only about 40 percent, but this does not mean that 60 percent of irrigation water is wasted. Some of the water not used by crops runs off the field or seeps back into groundwater and becomes a neighbor's supply. In these situations, improving efficiency may enhance water quality, leave more water in rivers for fisheries, and produce other important benefits, but it may also reduce supplies to other users elsewhere, thus failing to create real water savings.[91]

Studies of Egypt's Nile Valley, for instance, show that the irrigation efficiencies of individual farms are on the order of 40 percent, but that basinwide efficiency is already about 90 percent because of the multiple use of Nile water as it flows from the Aswan Dam to the Mediterranean Sea. In such cases, opportunities to save water may be limited to reducing evaporation losses and, where climate and soil conditions allow, switching to less water-intensive crops. This makes the water predicament of a country like Egypt even more difficult, since only minor additional water savings may be possible without removing land from irrigated agriculture.[92]

Water marketing can encourage efficiency and reuse.

Along with encouraging irrigation efficiency improvements, more appropriate pricing would help stretch the resource by promoting the treatment and reuse of urban wastewater for irrigation. This option is typically more expensive than most conservation and efficiency measures, but often less expensive than developing new water sources. Wastewater contains nitrogen and phosphorus, which can be pollutants when released to lakes and rivers, but are nutrients when applied to farmland. Moreover, unlike many other water sources, treated wastewater will be both an expanding and fairly reliable supply, since urban water use will likely double by 2025. Many large cities located along coastlines currently dump their wastewater, treated or untreated, into the ocean, rendering it unavailable for any other purpose and incidentally causing harm to coastal marine life. As long as the wastewater stream is kept free of heavy metals and harmful chemicals, and is treated adequately for irrigation use as far as disease vectors are concerned, it can become a vital new supply for agriculture.[93]

Israel, for example, reuses 65 percent of its domestic wastewater for crop production. At present, treated wastewater accounts for 30 percent of the nation's agricultural water supply, and this figure is expected to rise to 80 percent by 2025. Tunisia currently irrigates 3,000 hectares with treated wastewater and plans to increase this area tenfold by 2000. Worldwide, assuming domestic and municipal use doubles by 2025, the reuse of 65 percent of the resulting wastewater could boost agricultural supplies by 350 billion cubic meters per year—theoretically, enough to grow 350 million tons of wheat. Governments and development agencies, which are investing large sums in wastewater treatment plants, could promote this strategy by planning for agricultural reuse in the site selection, design, and construction of wastewater facilities.[94]

Along with more effective water pricing, water marketing can create incentives both to encourage efficiency and reuse, and to allocate water more productively. Instead of looking to a new dam or river diversion to get additional water, cities and farmers can purchase supplies from others who are willing to sell, trade, or lease their water or water rights. The Metropolitan Water District of Los Angeles, for example, is investing in conservation measures in southern California's Imperial Irrigation District in exchange for the water those investments will save. The annual cost of the conserved water is estimated at about 10¢ per cubic meter, far lower than the water district's best new-supply option. In Chile, where water policy directly encourages marketing, water companies that serve expanding cities now frequently buy small portions of water rights from farmers, most of whom have gained their surpluses through efficiency improvements.[95]

In most developing countries, water trading typically consists of spot sales or one-year lease arrangements, often between neighboring farmers. Weak institutions and difficulties with contract enforcement are frequent barriers to permanent sales of water rights. Although this limits marketing's potential benefits, the practice is still widespread. A 1990 survey of surface canal systems in Pakistan found active water trading in 70 percent of them. In India's western state of Gujarat, informal groundwater markets have emerged spontaneously and provide many farmers with water of high quality when needed, thus enhancing crop production. Since marketing may allow farmers who cannot afford to drill their own wells to purchase water from other irrigators, it can help provide the poor with access to irrigation water they otherwise would not have.[96]

Through markets, private organizations and government agencies can purchase existing water rights and dedicate them to restoring the aquatic environment. While private transactions cannot substitute for the government's responsibility in protecting freshwater ecosystems, they can play a useful supplementary role. In the western United States, there were 19 transactions reported during 1994 aimed at securing more water for rivers, wetlands, and other aquatic habitats. The Virginia-based Nature Conservancy, for example, has returned water to rivers and wetlands by outright purchases of private water rights as well as by working with state agencies to transfer existing water rights to instream uses. In Colorado, a coal-mining subsidiary of the Chevron Corporation donated $7.2 million worth of water rights on the Black Canyon of the Gunnison River to the Nature Conservancy, which then turned the rights over to the state Conservation Board for conversion to an instream water right. As a result, additional water will remain in this portion of the river to benefit three endangered fish species and a recreational trout fishery.[97]

Water marketing is not appropriate or workable everywhere, since it requires well-defined property rights to water.

TABLE 8

Projected Water Savings from U.S. Efficiency Standards, 1995–2025

	Fixture Water Use[1]		
	Without	With	
Year	Standards[2]	Standards[2]	Change
	(billion cubic meters per year)		(percent)
1995	25.8	25.2	-2
2000	25.7	24.0	-7
2010	25.2	20.8	-17
2020	24.3	16.9	-30
2025	25.2	16.1	-36

[1]Water use from toilets, faucets, and showerheads. [2]Standards as set in U.S. Energy Policy Act of 1992; water use declines slightly even without these standards because of less stringent standards already in place.
Source: Amy Vickers, Amy Vickers & Associates, Inc., *Technical Issues and Recommendations on the Implementation of the U.S. Energy Policy Act*, report prepared for the American Water Works Association, September 1995.

Moreover, if unregulated or monopolistic, water markets can lead to overexploitation of water sources, inequalities in water distribution, and exploitative prices. In India's southern state of Tamil Nadu, well owners pump groundwater, sometimes with the benefit of subsidized electricity, and sell it to intermediaries who in turn sell it to poor households lacking a piped water supply. The poor thus gain access to water, but may pay as much as ten times more for it than wealthier households connected to the public water system. A legal and regulatory framework is needed to protect the resource base and prevent one user group from unfairly exploiting another. Unless marketing takes place within such a framework, it can cause more harm than good.[98]

Efficiency standards round out the package of policy tools for stretching water sources. In the United States, legislation passed in late 1992 requires manufacturers of toilets, faucets, and showerheads to meet specified standards of efficiency as of January 1994. Today, the average U.S. resident's use of these fixtures takes an estimated 174 liters per day. But within 30 years, this figure is expected to drop by more than half—to 79 liters per day—as the more efficient models replace the existing stock. U.S. water utilities can thus plan on lower indoor water use, which translates into reduced capital investments for new water supplies and treatment plants, as well as reduced energy and chemical costs for pumping and treating water and wastewater. (See Table 8.) The environment benefits as well, since less water needs to be taken from rivers, lakes, and aquifers to meet urban needs, and since lowered energy use means fewer pollutants are emitted into the atmosphere.[99]

Along with the United States, a number of other governments, including Mexico and the Canadian province of Ontario, have adopted standards for household plumbing fixtures. The National Community Water Conservation Program in Cairo is currently working with the Egyptian government in attempts to introduce water conservation standards to the plumbing code there. Although efficiency standards have so far mainly been applied to household fixtures, they offer potential for water savings in agriculture, industry, and other municipal uses as well.[100]

Here and there, promising efforts inspire hope that the consequences of water shortage can at least be delayed. Yet, so far, concerted national and international efforts to bring all the pieces of a sustainable water strategy together are few. South Africa is worth watching in this regard. In early 1996, the Minister of Water Affairs and Forestry laid out principles for a fundamental overhaul of the nation's water law and management. Among the top priorities are providing each South African with access to at least 25 liters of water a day in order to meet the need for drinking water and sanitation; allocating water to the environment to prevent the loss of ecosystem functions, even if this requires taking water from other users; and reserving water for countries downstream in order to promote regional cooperation and integration. The principles also call for pricing water at levels that reflect its value, encouraging water marketing, and mandating that water suppliers adopt conservation measures. Although the principles are promising, turning them into actual laws, policies, and actions will not be easy because it involves dismantling decades of apartheid-era water legislation. Moreover, the country is still pursuing the socially and environmentally destructive Lesotho Highlands Water Development Project, an $8 billion dam-and-diversion scheme aimed at supplying the Johannesburg region with water from the tiny mountain kingdom of Lesotho. Nonetheless, the nation's new water law, which could be drafted by late 1996 and taken up by the parliament in early 1997, may emerge as one of the stronger national water strategies to date.[101]

At the international level, greater efforts are needed to assess and monitor the availability of water for food production on a worldwide basis. A basic blueprint for equitable and sustainable water use—which includes satisfying basic human and ecological needs, using and allocating water more efficiently, and sharing international waters equitably—will not guarantee agriculture the water supplies needed to meet the world's future food demands. Indeed, many of the policies and strategies to promote more sustainable water use—such as the raising of water prices and the expansion of water markets—will likely result in supplies being shifted away from agriculture. As Swedish hydrologist Malin Falkenmark and U.K.-based water specialist Asit Biswas note, "It would be erroneous to assume agriculture will maintain even its present total share of water use in the 21st century."[102]

The time may not be far off when a global grain bank will be needed to guard against food shortfalls induced by water shortages. Particularly in Africa, Asia, and the Middle East, water deficits will widen markedly in the coming decades. Together these regions are projected to grow by nearly 2.3 billion people by 2025; they account for 87 percent of projected population growth over the next 30 years. Many African and Asian countries are unlikely to have the financial resources to balance their water books by purchas-

ing surplus grain on the open market, assuming such surpluses exist. Thus, just as water security is being enhanced
in some regions by the establishment of water banks to cope
with droughts, food security may depend on the institution
of a global grain bank that can provide staple foods on con-
cessional terms for water-short, poor countries.[103]

Worldwide, population is expected to expand by 2.6
billion over the next 30 years, roughly the same number
that was added between 1950 and 1990, a period when global water use tripled. It seems highly unlikely that water use
can triple again: opportunities to capture more runoff
through dam construction are limited and groundwater in
many areas is already overtapped. Desalination will certainly expand, with the rate of growth dependent in large part
on energy prices and the pace of technological advances.
Seawater desalting with solar energy may come to occupy an
important supply niche, since many of the world's most
water-short regions are rich in sunshine. Yet desalination
capacity would have to expand 30-fold to supply even 5 percent of current world water use.

For the foreseeable future, reining in demand and dis-
tributing water more equitably—between people and
between nations, as well as between people and nature—
offer the best hope for preventing scarcity from leading to
more hunger and poverty, greater political and social instability, and more widespread ecological decline. Efficiency
gains can go a long way toward squeezing more out of the
existing supply. But water strategies alone will not be sufficient. Living within the limits of nature's water supply will
require reduced consumption among the more wealthy
social groups and reduced family size among all groups.
With nearly two out of every five tons of grain going into
meat and poultry production, individual choices about diet
collectively can influence how much water is needed to satisfy future food demands. And stepped-up efforts to create
the conditions needed for population stabilization must be
at the core of any successful strategy to achieve a sustainable
and secure water future for all.

By 2025—just a generation away—40 percent of the world's people may be living in countries experiencing water stress or chronic water scarcity. Successfully meeting the challenges of scarcity will require major changes in social institutions, economic policies, technologies, and ethics. It is too late for incremental progress to suffice. It is time to pull out all the stops.

Notes

1. Anita Alvarez de Williams, "Cocopa," in Alfonso Ortiz, ed., *Handbook of North American Indians*, vol. 10 (Washington, D.C.: Smithsonian, 1983); last nipa harvest from Edward Glenn and Richard Felger, "Saltwater Agriculture," *The World & I*, March 1994; author's communication with Cocopa community members, El Mayor, Baja California, Mexico, May 5, 1996.

2. Population projection from Population Reference Bureau (PRB), *1995 World Population Data Sheet* (Washington, D.C., 1995).

3. See United Nations, *Agenda 21: The United Nations Program of Action From Rio* (New York, 1992), and World Bank, *Water Resources Management: A World Bank Policy Paper* (Washington, D.C., 1993).

4. "Water-Use Targets Top Council's Tech Agenda," *IWRA Update* (newsletter of the International Water Resources Association), Summer 1996.

5. I. A. Shiklomanov, "World Fresh Water Resources," in Peter H. Gleick, ed., *Water in Crisis: A Guide to the World's Fresh Water Resources* (New York: Oxford University Press, 1993); renewable supply estimate from ranges in Gleick, op. cit. this note.

6. See Sandra L. Postel, Gretchen C. Daily, and Paul R. Ehrlich, "Human Appropriation of Renewable Fresh Water," *Science*, February 9, 1996.

7. Table 1 from Postel, Daily, and Ehrlich, op. cit. note 6, based on runoff estimates from M.I. L'Vovich et al., "Use and Transformation of Terrestrial Water Systems," in B.L. Turner et al., eds., *The Earth as Transformed by Human Action* (Cambridge: Cambridge University Press, 1990); Table 1 population estimates from PRB, op. cit. note 2; Amazon flow from E. Czaya, *Rivers of the World* (New York: Van Nostrand Reinhold, 1981); undammed northern rivers from Mats Dynesius and Christer Nilsson, "Fragmentation and Flow Regulation of River Systems in the Northern Third of the World," *Science*, November 4, 1994.

8. Distribution of runoff between flood flows and stable flows, and dam capacity figures from L'Vovich, op. cit. note 7.

9. Postel, Daily, and Ehrlich, op. cit. note 6.

10. Number of dams from Patrick McCully, *Silenced Rivers* (London: Zed Books, forthcoming); potential increase in accessible runoff from Postel, Daily, and Ehrlich, op. cit. note 6.

11. *Public Papers of the Presidents of the United States, John F. Kennedy, 1961* (Washington, D.C.: U.S. Government Printing Office, 1962).

12. December 1995 figures from Pat Burke, Secretary General, International Desalination Association, private communication, Topsfield, Mass., August 1, 1996; earlier figures from Wangnick Consulting, *1994 IDA Worldwide Desalting Plants Inventory*, Report No. 13 (Topsfield, Mass.: International Desalination Association, 1994), and Wangnick Consulting, *1990 IDA Worldwide Desalting Plants Inventory* (Englewood, N. J.: International Desalination Association, 1990).

13. Costs from World Bank, *From Scarcity to Security: Averting a Water Crisis in the Middle East and North Africa* (Washington, D.C., 1995), which gives a range of $1.00-1.50; completed in Santa Barbara, Calif., in 1992, the largest seawater desalination plant in the United States was estimated to produce water at $1.57 per cubic meter—it has since been mothballed; country populations from PRB, op. cit. note 2; energy requirements from Peter H. Gleick, "Energy and Water," *Annual Review of Energy and Environment*, Vol. 19, 1994.

14. Estimate of share consumed in agriculture from Shiklomanov, op. cit. note 5.

15. Figure of 1,000 tons from U.N. Food and Agriculture Organization (FAO), *Yield Response to Water* (Rome: 1979); calorie figures from Tim Dyson, *Population and Food: Global Trends and Future Prospects* (London: Routledge, 1996).

16. Worldwatch Institute estimate based on United States Department of Agriculture (USDA), Economic Research Service (ERS), "Production, Supply, and Distribution" (electronic database), Washington, D.C., August 1995, and on U.S. Bureau of the Census projections, published in Francis Urban and Ray Nightingale, *World Population by Country and Region, 1950-1990, with Projections to 2050* (Washington, D.C.: USDA, ERS, 1993).

17. FAO, op. cit. note 15; population projections from PRB, op. cit. note 2; annual flow of Nile from John Waterbury, *Hydropolitics of the Nile Valley* (Syracuse, N.Y.: Syracuse University Press, 1979); Colorado flow from U.S. Bureau of Reclamation, "Managing the Lower Colorado River to Meet Contemporary Needs," information sheet, Lower Colorado Region, Boulder City, Nev., n.d.

18. Gary Gardner, *Shrinking Fields: Cropland Loss in a World of Eight Billion* (Washington, D.C.: Worldwatch Institute, 1996); grainland area trend from Lester R. Brown, "Grain Area Unchanged," in Lester R. Brown, Hal Kane, and Ed Ayres, *Vital Signs 1993* (New York: W.W. Norton, 1993); irrigated area from FAO, *1990 Production Yearbook* (Rome, 1991), adjusted for the United States and Taiwan with irrigated area data from, respectively, (USDA), (ERS), *Agricultural Resources, Cropland, Water and Conservation*, Washington, D.C., September 1991, and Sophia Hung, USDA, ERS, private communication, June 21, 1991; harvest estimate of 40 percent is approximate, and is based on a 36 percent estimate in W. Robert Rangeley,

"Irrigation and Drainage in the World," in Wayne R. Jordan, ed., *Water and Water Policy in World Food Supplies* (College Station, Tex.: Texas A&M University Press, 1987); on a 47 percent estimate (just for grain) in Montague Yudelman, "The Future Role of Irrigation in Meeting the World's Food Supply," in Soil Science Society of America, *Soil and Water Science: Key to Understanding Our Global Environment* (Madison, Wisc., 1994); and on a general statement that 40 percent of world's food supply comes from irrigated land in Ismail Serageldin, *Toward Sustainable Management of Water Resources* (Washington, D.C.: World Bank, 1995).

19. Water stress and scarcity indicators from Malin Falkenmark and Carl Widstrand, "Population and Water Resources: A Delicate· Balance," *Population Bulletin*, PRB, Washington, D.C., 1992; Hillel Shuval, "Sustainable Water Resources Versus Concepts of Food Security, Water Security, Water Stress for Arid Countries," paper prepared for Stockholm Environment Institute/United Nations Workshop on Freshwater Resources, New York, May 18-19, 1996.

20. Middle East imports from Dyson, op. cit. note 15.

21. Postel, Daily, and Ehrlich, op. cit. note 6; in many developing countries, accessible runoff may be closer to 20 percent, according to Malin Falkenmark and Jan Lundqvist, "World Freshwater Problems: Call for a New Realism," background paper to the Comprehensive Global Water Resources Assessment, Stockholm Environment Institute, Stockholm, Sweden, 1996; minimum water requirement for urban and industrial needs from Shuval, op. cit. note 19.

22. Some of these countries benefit from rivers flowing in from neighbors, but these imported flows are not included so as to avoid double-counting. It should also be noted that some countries have substantial rainfall available for food production, even though they have only modest amounts of runoff. To the extent this rain falls on land suited for crop production, such countries can produce more food than the runoff figures alone would imply. In general, however, countries with relatively low runoff per person also have relatively low net precipitation per person. Grain imports from USDA, ERS, "Production, Supply, and Distribution" (electronic database), Washington, D.C., May 1996. Table 3 based on the following sources: Africa runoff figures from FAO, *Irrigation in Africa in Figures* (Rome: 1995); Middle East runoff figures from World Resources Institute (WRI), *World Resources 1994-95* (New York: Oxford University Press, 1994); population figures from PRB, op. cit. note 2; of the countries that would be added to the list in 2025, Botswana, Gambia, Senegal, and Swaziland would have more than 1,700 cubic meters per person if current inflow from other countries were included.

23. Population figures from PRB, op. cit. note 2; data on water-stressed countries from Robert Engelman and Pamela LeRoy, *Sustaining Water: Population and the Future of Renewable Water Supplies* (Washington, D.C.:

Population Action International, 1993); China's runoff from WRI, op. cit. note 22.

24. Table 4 drawn from the following sources: High Plains from Edwin D. Gutentag et al., *Geohydrology of the High Plains Aquifer in Parts of Colorado, Kansas, Nebraska, New Mexico, Oklahoma, South Dakota, Texas, and Wyoming* (Washington, D.C.: U.S. Government Printing Office, 1984); net extraction rates from Dork L. Sahagian, Frank W. Schwartz, and David K. Jacobs, "Direct Anthropogenic Contributions to Sea Level Rise in the Twentieth Century," *Nature*, January 6, 1994—this assumes an annual average depletion rate of 12 billion cubic meters from 1980-90 and adds it to the depletion estimate in Gutentag et al., op. cit. this note; irrigated area decline from Darrell S. Peckham and John B. Ashworth, *The High Plains Aquifer System of Texas, 1980 to 1990: Overview and Projections* (Austin, Tex.: Texas Water Development Board, 1993); California from California Department of Water Resources, *California Water Plan Update*, vol. 1 (Sacramento, Calif.: 1994); Southwest U.S. from Arizona Department of Water Resources, *Arizona Water Resources Assessment*, vols. 1 & 2 (Phoenix, 1994); T.W. Anderson et al., "Central Alluvial Basins," in W. Back, J.S. Rosenshein, and P.R. Seaber, eds., *Hydrogeology* (Boulder, Colo.: Geological Society of America, 1988); Albuquerque projection from "City's Conservation Plan on Target," *The Groundwater Newsletter*, Water Information Center, Inc., Denver, February 28, 1995; Mexico from Juan Manuel Martinez Garcia, Director General of Hydraulic Construction and Operation, Mexico City, private communication, October 21, 1991; Arabian Peninsula from Jamil Al Alawi and Mohammed Abdulrazzak, "Water in the Arabian Peninsula: Problems and Perspectives," in Peter Rogers and Peter Lydon, eds., *Water in the Arab World* (Cambridge: Harvard University, 1994); Abdulla Ali Al-Ibrahim, "Excessive Use of Groundwater Resources in Saudi Arabia: Impacts and Policy Options," *Ambio*, February 1991; Libya calculated from Rajab M. El Asswad, "Agricultural Prospects and Water Resources in Libya," *Ambio*, Vol. 24, No. 6, September 1995; African Sahara from Sahagian, Schwartz, and Jacobs, op. cit. this note; Israel, Gaza, and Spain from I. A. Shiklomanov, "Assessment of Water Resources and Water Availability in the World," State Hydrological Institute, St. Petersburg, Russia, February 1996; India's Punjab from International Rice Research Institute (IRRI), *Water: A Looming Crisis* (Manila, 1995); A. Vaidyanathan, "Second India Series Revisited: Food and Agriculture," report prepared for WRI, Washington, D.C., n.d.; Harald Frederiksen, Jeremy Berkoff, and William Barber, *Water Resources Management in Asia* (Washington, D.C.: World Bank, 1993); China from Xu Zhifang, unpublished paper prepared for World Water Council-Interim Founding Committee, March 1995; Southeast Asia from Frederiksen, Berkoff, and Barber, op. cit. this note.

25. Depletion figure from Peter H. Gleick et al., *California Water 2020: A Sustainable Vision* (Oakland, Calif.: Pacific Institute for Studies in Development, Environment, and Security, 1995); cost of storage projects from California Department of Water Resources, op. cit. note 24.

26. Ganges from Frederiksen, Berkoff, and Barber, op. cit. note 24; drying of Yellow River from Patrick E. Tyler, "China's Fickle Rivers: Dry Farms, Needy Industry Bring a Water Crisis," *New York Times*, May 23, 1996.

27. Sandra Postel, "Irrigation Expansion Slowing," in Lester R. Brown, Hal Kane, and David Malin Roodman, *Vital Signs 1994* (New York: W.W. Norton, 1994); Gary Gardner, "Irrigated Area Dips Slightly," in Lester R. Brown, Christopher Flavin, and Hal Kane, *Vital Signs 1996* (New York: W.W. Norton, 1996); per capita grain production from Lester R. Brown, "World Grain Production Falls," in Lester R. Brown, Christopher Flavin, and Hal Kane, *Vital Signs 1996* (New York: W.W. Norton, 1996); irrigated area data for Figure 1 from FAO, *Production Yearbook* (Rome, various years), and Bill Quimby, USDA, ERS, personal communication, January 1996.

28. Sandra Postel, *Last Oasis: Facing Water Scarcity* (New York: W.W. Norton, 1992); Dina L. Umali, in *Irrigation-Induced Salinity* (Washington, D.C.: World Bank, 1993), cites sources suggesting that 2-3 million hectares a year may be coming out of production due to salinization, which, if accurate, would offset the 2 million hectares of average annual irrigation expansion in recent years; David Seckler, *The New Era of Water Resources Management: From 'Dry' to 'Wet' Water Savings* (Washington, D.C.: Consultative Group on International Agricultural Research, 1996).

29. Urbanization in 2025 from Gershon Feder and Andrew Keck, "Increasing Competition for Land and Water Resources: A Global Perspective," paper prepared for World Bank, Washington, D.C., March 1995.

30. 1957 figures from Gleick et al., op. cit. note 25; California Department of Water Resources, op. cit. note 24.

31. Frederiksen, Berkoff, and Barber, op. cit. note 24; 300 cities from Xu, op. cit. note 24; Patrick E. Tyler, "China Lacks Water to Meet Its Mighty Thirst," *New York Times*, November 7, 1993.

32. Philip Shenon, "Fore! Golf in Asia Hits Environmental Rough," *New York Times*, October 22, 1994; see also Anne E. Platt, "Toxic Green: The Trouble with Golf," *World Watch*, May/June 1994.

33. Klaus Lampe, "'...Our Daily Bread,'" *Swiss Review of World Affairs*, September 1994.

34. Paul J. Kramer and John S. Boyer, *Water Relations of Plants and Soils* (San Diego, Calif: Academic Press, 1995).

35. IRRI, op. cit. note 24.

36. Israeli examples from A. Benin, "Utilization of Recycled, Saline and other Marginal Waters for Irrigation: Challenges and Management Issues," in F. Lopez-Vera, J. De Castro Morcillo, and A. Lopez Lillio, eds., *Uso del Agua en las Areas Verdes Urbanas* (Madrid, 1993); California example from Seckler, op. cit. note 28; Edward P. Glenn et al., *"Salicornia bigelovii* Torr.: An Oilseed Halophyte for Seawater Irrigation," *Science*, March 1, 1991; U.S. National Research Council, *Saline Agriculture: Salt-Tolerant Plants for Developing Countries* (Washington, D.C.: National Academy Press, 1990).

37. Author's visit to and overflight of the Colorado Delta, May 1996; Aldo Leopold, *A Sand County Almanac* (New York: Oxford University Press, 1949).

38. Sandra Postel, "Where Have All the Rivers Gone?" *World Watch*, May/June 1995; in Figure 2, 1905-49 is flow at Yuma, Ariz., with data provided by the U.S. Geological Survey, Denver; 1950-92 is flow at southerly international boundary, with data provided by the International Boundary and Water Commission, El Paso, Tex.

39. Global demand from Postel, op. cit. note 28; Jan A. Veltrop, "Importance of Dams for Water Supply and Hydropower," in Asit K. Biswas, Mohammed Jellali, and Glenn Stout, *Water for Sustainable Development in the 21st Century* (Oxford: Oxford University Press, 1993); current number of dams from Patrick McCully, International Rivers Network, Berkeley, Calif., private communication, February 1995.

40. Sandra Postel and Stephen Carpenter, "Freshwater Ecosystem Services," in Gretchen C. Daily, ed., *Nature's Services: Societal Dependence on Natural Ecosystems* (Washington, D.C.: Island Press, forthcoming).

41. The Nature Conservancy, *Priorities for Conservation: 1996 Annual Report Card for U.S. Plant and Animal Species* (Arlington, Va., 1996).

42. Gleick et al., op. cit. note 25.

43. Frederiksen, Berkoff, and Barber, op. cit. note 24.

44. Khalil H. Mancy, "The Environmental and Ecological Impacts of the Aswan High Dam," in Hillel Shuval, ed., *Developments in Arid Zone Ecology and Environmental Quality* (Philadelphia, Pa.: Balaban ISS, 1981); Gilbert White, "The Environmental Effects of the High Dam at Aswan," *Environment*, September 1988.

45. Fred Pearce, "High and Dry in Aswan," *New Scientist*, May 7, 1994.

46. Ibid.; John D. Milliman, James M. Broadus, and Frank Gable, "Environmental and Economic Implications of Rising Sea Level, *Ambio*, Vol. 18, No. 6, 1989.

47. Philip P. Micklin, "The Water Management Crisis in Soviet Central Asia," *The Carl Beck Papers in Russian and East European Studies* (Pittsburgh, Pa.: University of Pittsburgh, 1991); basin's irrigated area from Aral Sea

Basin Program-Group 1, Interstate Commission for Water Coordination, and World Bank, "Developing a Regional Water Management Strategy: Issues and Work Plan," draft prepared for the Executive Committee of the Interstate Council for the Aral Sea, May 1996.

48. Philip Micklin, "The Aral Crisis: Introduction to the Special Issue," *Post-Soviet Geography*, May 1992; Figure 3 from Philip Micklin, as published in Gleick, ed., op. cit. note 5; inflow in recent years from Aral Sea Basin Program-Group 1 et al., op. cit. note 47.

49. Loss of fish species from Judith Perera, "A Sea Turns to Dust," *New Scientist*, October 23, 1993; Philip Micklin, "Touring the Aral: Visit to an Ecological Disaster Zone," *Soviet Geography*, February 1991.

50. Micklin, op. cit. note 48.

51. Owen Lammers and Lillian Phaeton, "Hidrovia: Will A New Transport Route Destroy Brazil's Pantanal?" *World Rivers Review*, First Quarter 1994; "Environmental Concerns May Doom Bank Financing for Waterway Project," *International Environment Reporter*, June 28, 1995; "Brazil: The Waterway," *Economist*, February 17, 1996; number of species from Alan P. Covich, "Water and Ecosystems," in Gleick, ed., op. cit. note 5; Diana Jean Schemo, "Ecologists Criticize First Steps in Latin River Plan," *New York Times*, May 26, 1996; Owen Lammers, International Rivers Network, private communication, Berkeley, Calif., August 9, 1996.

52. Elizabeth Culotta, "Bringing Back the Everglades," *Science*, Vol. 268, June 23, 1995; Norman Boucher, "Back to the Everglades," *Technology Review*, August-September 1995.

53. Thomas F. Homer-Dixon, "Environmental Scarcities and Violent Conflict," *International Security*, Summer 1994.

54. Ibid.; number of wells and water prices from Jad Isaac and Jan Selby, "The Palestinian Water Crisis," *Natural Resources Forum*, Vol. 20, No. 1, 1996; swimming pools from David A. Schwarzbach, "Promised Land. (But What About the Water?)," *The Amicus Journal*, Summer 1995; see also Miriam R. Lowi, "West Bank Water Resources and the Resolution of Conflict in the Middle East," Occasional Paper Series, Project on Environmental Change and Acute Conflict, September 1992; and Information Division, "Israeli-Palestinian Interim Agreement Annex III—Protocol Concerning Civil Affairs," Israeli Foreign Ministry, Jerusalem, September 1995.

55. Information Division, op. cit. note 54; "Water in the Middle East: As Thick as Blood," *Economist*, December 23, 1995-January 5, 1996; Isaac and Selby, op. cit. note 54.

56. Homer-Dixon, op. cit. note 53.

57. Michael Goldman, "Tragedy of the Commons or the Commoners' Tragedy: The State and Ecological Crisis in India," *CNS*, December 1993.

58. Homer-Dixon, op. cit. note 53; Table 6 based on the following sources: Turkmenistan and Uzbekistan figures from David R. Smith, "Climate Change, Water Supply, and Conflict in the Aral Sea Basin," paper presented at the "PriAral Workshop 1994," San Diego State University, March 1994; others from Gleick, ed., op. cit. note 5.

59. Thomas Naff, "Conflict and Water Use in the Middle East," in Rogers and Lydon, eds., op. cit. note 24; perspective on Six Day War from Daniel Hillel, *Rivers of Eden: The Struggle for Water and the Quest for Peace in the Middle East* (New York: Oxford University Press, 1994).

60. Naff, op. cit. note 59; Egypt irrigated area from FAO, op. cit. note 22.

61. Z. Abate, "The Integrated Development of Nile Basin Waters," in P.P. Howell and J.A. Allan, eds., *The Nile: Sharing a Scarce Resource* (Cambridge: Cambridge University Press, 1994); Uganda figure based on estimated additional withdrawals of 11 million cubic meters per day to satisfy full irrigation potential, in Malin Falkenmark and Jan Lundqvist, "Looming Water Crisis: New Approaches are Inevitable," in Leif Ohlsson, ed., *Hydropolitics: Conflicts Over Water as a Development Constraint* (London: Zed Books, 1995).

62. Author's visit to the region, March 1995; examples of local-regional conflict from Smith, op. cit. note 58; "Central Asia: The Silk Road Catches Fire," *Economist*, December 26, 1992-January 8, 1993.

63. Postel, op. cit. note 28.

64. Mahbubul Alam, "Sacred Ganges Becomes River of Woe," *The WorldPaper*, November 1994; Sheila Jones, "When the Ganges Runs Dry," *Financial Times*, May 9, 1994; Gordon Platt, "India's Control of Ganges River Flow a 'Life and Death' Issue for Bangladesh," *Journal of Commerce*, October 26, 1995.

65. Sandra Postel, "The Politics of Water," *World Watch*, July/August 1993; water-sharing between Syria and Iraq from John Kolars, "River Advocacy and Return Flow Management on the Euphrates/Firat River: An Important Element in Core-Periphery Relations," prepared for the conference, *Water: A Trigger for Conflict/A Reason for Cooperation*, Indiana Center on Global Change and World Peace, Bloomington, Ind., March 7-10, 1996.

66. Quote reported in John Murray Brown, "Turkey, Syria Set Talks on Euphrates," *Washington Post*, January 22, 1993; Manuel Schiffler, report on the "Interdisciplinary Academic Conference on Water in the Middle East," German Development Institute, Berlin, June 17-18, 1995.

67. Shared rivers from Asit K. Biswas, "Management of International Water

Resources: Some Recent Developments," in Asit K. Biswas, ed., *International Waters of the Middle East* (Oxford: Oxford University Press, 1994), who points out that the actual number of international rivers must be higher than the 214 estimated by a now-defunct United Nations agency.

68. For a good discussion of international water law, see Stephen C. McCaffrey, "Water, Politics, and International Law," in Gleick, ed., op. cit. note 5; Gail Bingham, Aaron Wolf, and Tim Wohlgenant, *Resolving Water Disputes: Conflict and Cooperation in the United States, the Near East, and Asia* (Washington, D.C.: U.S. Agency for International Development, 1994).

69. Courtney G. Flint, "Recent Developments of the International Law Commission Regarding International Watercourses and Their Implications for the Nile River," *Water International*, Vol. 20, No. 4, 1995.

70. McCaffrey, op. cit. note 68.

71. Warren Christopher, "American Diplomacy and the Global Environmental Challenges of the 21st Century," text of speech given at Stanford University, Stanford, Calif., April 9, 1996.

72. "The Nile Basin Initiative," *Water and Sustainable Development*, Canadian International Development Agency, March 1996; "Extracts from the Minutes of the 3rd Meeting of the Ministers of Water Affairs in the Nile Basin on Tecconile," and Annex Z, "Project on the Nile Basin Cooperative Framework, Draft Terms of Reference for a Panel of Experts Constituted by the Tecconile Council of Ministers," Arusha, Tanzania, February 9-11, 1995.

73. Aral Sea Basin Program-Group 1 et al., op. cit. note 47.

74. Ibid.

75. Peter Rogers, "The Value of Cooperation in Resolving International River Basin Disputes," *Natural Resources Forum*, May 1993; refugee movement from Jones, op. cit. note 64.

76. Syed S. Kirmani, "Water, Peace and Conflict Management: The Experience of the Indus and Mekong River Basins," *Water International*, Vol. 15, No. 4, 1990; Jagat S. Mehta, "The Indus Water Treaty: A Case Study in the Resolution of an International River Basin Conflict," *Natural Resources Forum*, Vol. 12, No. 1, 1988.

77. Sander Thoenes, "Central Asians Reach Common Ground over Water," *Financial Times*, April 9, 1996; Ganges example is that of George Verghese of the Centre for Policy Research in New Delhi, as cited in John Zubrzycki, "Where Water is not their Own," *Christian Science Monitor*, March 27, 1996.

78. Bingham, Wolf, and Wohlgenant, op. cit. note 68.

79. Franklin Fisher, "The Economics of Water Dispute Resolution, Project Evaluation and Management: An Application to the Middle East," paper presented at Stockholm Water Symposium, Sweden, August 14, 1995.

80. D.J. Blackmore, "Integrated Catchment Management—The Murray-Darling Basin Experience," paper presented at "Water Down Under '94," Adelaide, Australia, November 21-25, 1994; Murray-Darling Basin Commission, *The Murray-Darling Basin Commission: Managing Australia's Heartland* (Canberra, 1992).

81. J.A. Allan, "The Nile Basin: Water Management Strategies," in Howell and Allan, eds., op. cit. note 61; Arab Sea Basin Program-Group 1 et al., op. cit. note 47.

82. Peter H. Gleick, "Basic Water Requirements for Human Activities: Meeting Basic Needs," *Water International*, June 1996; see also Ashok Nigam and Gourisankar Ghosh, "A Model of Costs and Resources for Rural and Peri-Urban Water Supply and Sanitation in the 1990s," *Water Resources Journal*, March 1996.

83. Brian Gray, "The Modern Era in California Water Law," *Hastings Law Journal*, January 1994; John H. Cushman, Jr., "U.S. and California Sign Water Accord," *New York Times*, December 16, 1994; Daniel Sneider, "Mono Lake's Resurrection Is a Model for Watershed Battles," *Christian Science Monitor*, August 1, 1996; Postel, op. cit. note 28.

84. Aral Sea Program Unit, "Aral Sea Program—Phase 1," World Bank, Washington, D.C., May 1994; N.F. Glazovskiy, "Ideas on an Escape from the 'Aral Crisis,'" *Soviet Geography*, February 1991.

85. Mangrove example from Janet N. Abramovitz, *Imperiled Waters, Impoverished Future: The Decline of Freshwater Ecosystems* (Washington, D.C.: Worldwatch Institute, 1996); western United States from Postel and Carpenter, op. cit. note 40.

86. World Bank, *From Scarcity to Security: Averting a Water Crisis in the Middle East and North Africa* (Washington, D.C., 1995); for more extensive discussion of efficiency measures, see Postel, op. cit note 28.

87. Tunisia and Jordan examples from World Bank, op. cit. note 86; Richard W. Wahl, *Markets for Federal Water: Subsidies, Property Rights, and the Bureau of Reclamation* (Washington, D.C.: Resources for the Future, 1989); also see Subcommittee on Oversight and Investigations, Committee on Natural Resources, U.S. House of Representatives, *Taking from the Taxpayer: Public Subsidies for Natural Resource Development* (Washington, D.C., 1994), which cites estimates of the total irrigation subsidy of $34-70 billion; India figure from Vaidyanathan, op. cit. note 24.

88. Described in Gleick et al., op. cit. note 25.

89. Postel, op. cit. note 28.

90. Willem Van Tuijl, *Improving Water Use in Agriculture: Experiences in the Middle East and North Africa* (Washington, D.C.: World Bank, 1993); Paul Polak, "Progress Report on IDE Low Cost Drip Irrigation System," prepared for International Development Enterprises, Lakewood, Colo., April 9, 1995; International Development Enterprises, "Tripling the Harvest of Small Farmers," video production, Lakewood, Colo.

91. Worldwide irrigation efficiency from Postel, op. cit. 28.

92. Andrew A. Keller and Jack Keller, "Effective Efficiency: A Water Use Efficiency Concept for Allocating Freshwater Resources," paper prepared for Center for Economic Policy Studies, Winrock International, Arlington, Va., 1995; irrigation expansion from Hussam Fahmy, "Comparative Analysis of Egyptian Water Policies," *Water International*, Vol. 21, No. 1, March 1996.

93. Benin, op. cit. note 36, observes that because treatment typically does not remove the salts from wastewater, reuse may hasten salinization of soils over the long term.

94. Israel's reuse from Hillel Shuval, presentation at the Stockholm Environment Institute/United Nations Workshop on Freshwater Resources, New York, May 18-19, 1996; worldwide municipal use from Shiklomanov, op. cit. note 5; Tunisia from World Bank, op. cit. note 86; worldwide calculation assumes that 10 percent of domestic use is consumed, leaving 90 percent as wastewater; it also assumes that 1,000 cubic meters of water are needed to produce a ton of wheat, including only direct evapotranspiration requirements.

95. Postel, op. cit. note 28; Renato Gazmuri Schleyer and Mark W. Rosegrant, "Chilean Water Policy: The Role of Water Rights, Institutions, and Markets," in Mark W. Rosegrant and Renato Gazmuri Schleyer, *Tradable Water Rights: Experiences in Reforming Water Allocation Policy* (Arlington, Va.: Irrigation Support Project for Asia and the Near East, 1994).

96. Mateen Thobani, "Tradable Property Rights to Water," FPD Note No. 34, World Bank, Washington, D.C., February 1995; Tushaar Shah, *Groundwater Markets and Irrigation Development* (Bombay: Oxford University Press, 1993).

97. Western water transactions from Rodney T. Smith and Roger Vaughan, "1994 Annual Transaction Review: Markets Expanding to New Areas," *Water Strategist* (Stratecon, Inc., Claremont, Calif.: January 1995); Robert Wigington, "Market Strategies for the Protection of Western Instream Flows and Wetlands," Natural Resources Law Center, University of Colorado School of Law, Boulder, August 1990; Peter Steinhart, "The Water Profiteers," *Audubon*, March 1990.

98. India example from Kuppannan Palanisami, "Evolution of Agricultural and Urban Water Markets in Tamil Nadu, India," in Rosegrant and Schleyer, op. cit. note 95.

99. Amy Vickers, "The Energy Policy Act: Assessing Its Impact on Utilities," *Journal AWWA* (Journal of the American Water Works Association), August 1993; see also Janice A. Beecher, "Integrated Resource Planning," *Journal AWWA*, June 1995.

100. Mexico and Ontario from Postel, op. cit. note 28; Cairo from "Waterwiser," American Water Works Association, World Wide Web information clearinghouse, May 30, 1996.

101. "Water Law Review Supported By All," draft press release, Ministry of Water Affairs and Forestry, Pretoria, South Africa, February 6, 1996; "A Step Towards Equity in Access and Optimal Use of Water," *Water Sewage and Effluent*, Vol. 16, No. 1, March 1996; Eddie Koch, "A Watershed for Apartheid," *New Scientist*, April 13, 1996; Korinna Horta, "Making the Earth Rumble: The Lesotho-South African Water Connection," *Multinational Monitor*, May 1996.

102. Malin Falkenmark and Asit K. Biswas, "Further Momentum to Water Issues," *Ambio*, Vol. 24, No. 6, 1995.

103. PRB, op. cit. note 2.

PUBLICATION ORDER FORM

No. of
Copies

_____ 68. **Banishing Tobacco** by William U. Chandler.
_____ 70. **Electricity For A Developing World: New Directions** by Christopher Flavin.
_____ 75. **Reassessing Nuclear Power: The Fallout From Chernobyl** by Christopher Flavin.
_____ 77. **The Future of Urbanization: Facing the Ecological and Economic Constraints**
         by Lester R. Brown and Jodi L. Jacobson.
_____ 78. **On the Brink of Extinction: Conserving The Diversity of Life** by Edward C. Wolf.
_____ 79. **Defusing the Toxics Threat: Controlling Pesticides and Industrial Waste**
         by Sandra Postel.
_____ 80. **Planning the Global Family** by Jodi L. Jacobson.
_____ 81. **Renewable Energy: Today's Contribution, Tomorrow's Promise** by
         Cynthia Pollock Shea.
_____ 82. **Building on Success: The Age of Energy Efficiency** by Christopher Flavin
         and Alan B. Durning.
_____ 84. **Rethinking the Role of the Automobile** by Michael Renner.
_____ 86. **Environmental Refugees: A Yardstick of Habitability** by Jodi L. Jacobson.
_____ 89. **National Security: The Economic and Environmental Dimensions** by Michael Renner.
_____ 90. **The Bicycle: Vehicle for a Small Planet** by Marcia D. Lowe.
_____ 91. **Slowing Global Warming: A Worldwide Strategy** by Christopher Flavin
_____ 92. **Poverty and the Environment: Reversing the Downward Spiral** by Alan B. Durning.
_____ 93. **Water for Agriculture: Facing the Limits** by Sandra Postel.
_____ 94. **Clearing the Air: A Global Agenda** by Hilary F. French.
_____ 95. **Apartheid's Environmental Toll** by Alan B. Durning.
_____ 96. **Swords Into Plowshares: Converting to a Peace Economy** by Michael Renner.
_____ 97. **The Global Politics of Abortion** by Jodi L. Jacobson.
_____ 98. **Alternatives to the Automobile: Transport for Livable Cities** by Marcia D. Lowe.
_____ 99. **Green Revolutions: Environmental Reconstruction in Eastern Europe and the**
         Soviet Union by Hilary F. French.
_____ 100. **Beyond the Petroleum Age: Designing a Solar Economy** by Christopher Flavin
         and Nicholas Lenssen.
_____ 101. **Discarding the Throwaway Society** by John E. Young.
_____ 102. **Women's Reproductive Health: The Silent Emergency** by Jodi L. Jacobson.
_____ 103. **Taking Stock: Animal Farming and the Environment** by Alan B. Durning and
         Holly B. Brough.
_____ 104. **Jobs in a Sustainable Economy** by Michael Renner.
_____ 105. **Shaping Cities: The Environmental and Human Dimensions** by Marcia D. Lowe.
_____ 106. **Nuclear Waste: The Problem That Won't Go Away** by Nicholas Lenssen.
_____ 107. **After the Earth Summit: The Future of Environmental Governance**
         by Hilary F. French.
_____ 108. **Life Support: Conserving Biological Diversity** by John C. Ryan.
_____ 109. **Mining the Earth** by John E. Young.
_____ 110. **Gender Bias: Roadblock to Sustainable Development** by Jodi L. Jacobson.
_____ 111. **Empowering Development: The New Energy Equation** by Nicholas Lenssen.
_____ 112. **Guardians of the Land: Indigenous Peoples and the Health of the Earth**
         by Alan Thein Durning.
_____ 113. **Costly Tradeoffs: Reconciling Trade and the Environment** by Hilary F. French.
_____ 114. **Critical Juncture: The Future of Peacekeeping** by Michael Renner.
_____ 115. **Global Network: Computers in a Sustainable Society** by John E. Young.
_____ 116. **Abandoned Seas: Reversing the Decline of the Oceans** by Peter Weber.
_____ 117. **Saving the Forests: What Will It Take?** by Alan Thein Durning.
_____ 118. **Back on Track: The Global Rail Revival** by Marcia D. Lowe.
_____ 119. **Powering the Future: Blueprint for a Sustainable Electricity Industry**
         by Christopher Flavin and Nicholas Lenssen.
_____ 120. **Net Loss: Fish, Jobs, and the Marine Environment** by Peter Weber.
_____ 121. **The Next Efficiency Revolution: Creating a Sustainable Materials Economy**
         by John E. Young and Aaron Sachs.

_____ **Total Copies**

Single Copy: $5.00 • 2–5: $4.00 ea. • 6–20: $3.00 ea. • 21 or more: $2.00 ea.
Call Director of Communication at (202) 452-1999 to inquire about discounts on larger orders.

☐ **Membership in the Worldwatch Library: $30.00 (international airmail $45.00)**
The paperback edition of our 250-page "annual physical of the planet,"
State of the World, plus all Worldwatch Papers released during the calendar year.

☐ **Subscription to *World Watch* magazine: $20.00 (international airmail $35.00)**
Stay abreast of global environmental trends and issues with our award-winning, eminently readable bimonthly magazine.

☐ **Worldwatch Database Disk Subscription: One year for $89**
Includes current global agricultural, energy, economic, environmental, social, and military indicators from all current Worldwatch publications. Includes a mid-year update, and *Vital Signs* and *State of the World* as they are published. Can be used with Lotus 1-2-3, Quattro Pro, Excel, SuperCalc and many other spreadsheets.
Check one: _____high-density IBM-compatible or _____Macintosh

---

Make check payable to Worldwatch Institute
1776 Massachusetts Avenue, N.W., Washington, D.C. 20036-1904 USA

Please include $3 postage and handling for non-subscription orders.

Enclosed is my check for U.S. $_____
AMEX ☐   VISA ☐   Mastercard ☐   _____
                                  Card Number   Expiration Date

name   **daytime phone #**

address

city   **state**   **zip/country**

Phone: (202) 452-1999   Fax: (202) 296-7365   E-Mail: wwpub@worldwatch.org   WWP

PUBLICATION ORDER FORM

No. of
Copies

_____ 68. **Banishing Tobacco** by William U. Chandler.

_____ 70. **Electricity For A Developing World: New Directions** by Christopher Flavin.

_____ 75. **Reassessing Nuclear Power: The Fallout From Chernobyl** by Christopher Flavin.

_____ 77. **The Future of Urbanization: Facing the Ecological and Economic Constraints**
by Lester R. Brown and Jodi L. Jacobson.

_____ 78. **On the Brink of Extinction: Conserving The Diversity of Life** by Edward C. Wolf.

_____ 79. **Defusing the Toxics Threat: Controlling Pesticides and Industrial Waste**
by Sandra Postel.

_____ 80. **Planning the Global Family** by Jodi L. Jacobson.

_____ 81. **Renewable Energy: Today's Contribution, Tomorrow's Promise** by
Cynthia Pollock Shea.

_____ 82. **Building on Success: The Age of Energy Efficiency** by Christopher Flavin
and Alan B. Durning.

_____ 84. **Rethinking the Role of the Automobile** by Michael Renner.

_____ 86. **Environmental Refugees: A Yardstick of Habitability** by Jodi L. Jacobson.

_____ 89. **National Security: The Economic and Environmental Dimensions** by Michael Renner.

_____ 90. **The Bicycle: Vehicle for a Small Planet** by Marcia D. Lowe.

_____ 91. **Slowing Global Warming: A Worldwide Strategy** by Christopher Flavin

_____ 92. **Poverty and the Environment: Reversing the Downward Spiral** by Alan B. Durning.

_____ 93. **Water for Agriculture: Facing the Limits** by Sandra Postel.

_____ 94. **Clearing the Air: A Global Agenda** by Hilary F. French.

_____ 95. **Apartheid's Environmental Toll** by Alan B. Durning.

_____ 96. **Swords Into Plowshares: Converting to a Peace Economy** by Michael Renner.

_____ 97. **The Global Politics of Abortion** by Jodi L. Jacobson.

_____ 98. **Alternatives to the Automobile: Transport for Livable Cities** by Marcia D. Lowe.

_____ 99. **Green Revolutions: Environmental Reconstruction in Eastern Europe and the
Soviet Union** by Hilary F. French.

_____100. **Beyond the Petroleum Age: Designing a Solar Economy** by Christopher Flavin
and Nicholas Lenssen.

_____101. **Discarding the Throwaway Society** by John E. Young.

_____102. **Women's Reproductive Health: The Silent Emergency** by Jodi L. Jacobson.

_____103. **Taking Stock: Animal Farming and the Environment** by Alan B. Durning and
Holly B. Brough.

_____104. **Jobs in a Sustainable Economy** by Michael Renner.

_____105. **Shaping Cities: The Environmental and Human Dimensions** by Marcia D. Lowe.

_____106. **Nuclear Waste: The Problem That Won't Go Away** by Nicholas Lenssen.

_____107. **After the Earth Summit: The Future of Environmental Governance**
by Hilary F. French.

_____108. **Life Support: Conserving Biological Diversity** by John C. Ryan.

_____109. **Mining the Earth** by John E. Young.

_____110. **Gender Bias: Roadblock to Sustainable Development** by Jodi L. Jacobson.

_____111. **Empowering Development: The New Energy Equation** by Nicholas Lenssen.

_____112. **Guardians of the Land: Indigenous Peoples and the Health of the Earth**
by Alan Thein Durning.

_____113. **Costly Tradeoffs: Reconciling Trade and the Environment** by Hilary F. French.

_____114. **Critical Juncture: The Future of Peacekeeping** by Michael Renner.

_____115. **Global Network: Computers in a Sustainable Society** by John E. Young.

_____116. **Abandoned Seas: Reversing the Decline of the Oceans** by Peter Weber.

_____117. **Saving the Forests: What Will It Take?** by Alan Thein Durning.

_____118. **Back on Track: The Global Rail Revival** by Marcia D. Lowe.

_____119. **Powering the Future: Blueprint for a Sustainable Electricity Industry**
by Christopher Flavin and Nicholas Lenssen.

_____120. **Net Loss: Fish, Jobs, and the Marine Environment** by Peter Weber.

_____121. **The Next Efficiency Revolution: Creating a Sustainable Materials Economy**
by John E. Young and Aaron Sachs.

_____ **Total Copies**

Single Copy: $5.00 • 2–5: $4.00 ea. • 6–20: $3.00 ea. • 21 or more: $2.00 ea.
Call Director of Communication at (202) 452-1999 to inquire about discounts on larger orders.

☐ **Membership in the Worldwatch Library: $30.00 (international airmail $45.00)**
    The paperback edition of our 250-page "annual physical of the planet,"
    State of the World, plus all Worldwatch Papers released during the calendar year.

☐ **Subscription to *World Watch* magazine: $20.00 (international airmail $35.00)**
    Stay abreast of global environmental trends and issues with our award-winning,
    eminently readable bimonthly magazine.

☐ **Worldwatch Database Disk Subscription: One year for $89**
Includes current global agricultural, energy, economic, environmental, social, and military indicators from all current Worldwatch publications. Includes a mid-year update; and *Vital Signs* and *State of the World* as they are published. Can be used with Lotus 1-2-3, Quattro Pro, Excel, SuperCalc and many other spreadsheets.
Check one: _____high-density IBM-compatible or _____Macintosh

---

Make check payable to Worldwatch Institute
1776 Massachusetts Avenue, N.W., Washington, D.C. 20036-1904 USA

Please include $3 postage and handling for non-subscription orders.

Enclosed is my check for U.S. $_____
AMEX☐ VISA☐ Mastercard☐ _____
                                   Card Number   Expiration Date

name   **daytime phone #**

address

city   **state**   **zip/country**

Phone: (202) 452-1999   Fax: (202) 296-7365   E-Mail: wwpub@worldwatch.org WWP